T0173072

Water Purification Using Heat Pumps

Water Purification Using Heat Pumps

F. A. Holland
Overseas Educational Development Office, University of Salford, UK,
and Universidad Autónoma del Estado de Morelos UAEM, Mexico.

J. Siqueiros
Universidad Autónoma del Estado de Morelos UAEM, Mexico.

S. Santoyo
Instituto de Investigaciones Eléctricas IIE, Mexico.

C. L. Heard
Instituto Mexicano del Petróleo IMP, Mexico.

and

E. R. Santoyo
Universidad Nacional Autónoma de México UNAM, Mexico.

First published 1999
by E & FN Spon
11 New Fetter Lane, London EC4P 4EE

Simultaneously published in the USA and Canada
by Routledge
29 West 35th Street, New York, NY 10001

E & FN Spon is an imprint of the Taylor & Francis Group

Typeset in 10/12pt Sabon by Puretech India Ltd, Pondicherry
http://www.puretech.com.
Printed and bound in Great Britain by TJ International Ltd, Padstow, Cornwall

British Library Cataloguing in Publication Data
A catalogue record for this book is available from the British Library

Library of Congress Cataloging in Publication Data
Water purification using heat pumps / F. A. Holland ... [et al.].
    p. cm.
  Includes bibliographical references and index.
  (hardbound : alk. paper)
    1. Water–Purification. 2. Heat pumps. I. Holland, F. A.
TD430.W3647   1999
628.1'64–dc21   98-51598
                                                        CIP

ISBN 0–419–24710–6

Contents

6 ECONOMICS OF HEAT PUMP SYSTEMS

Foreword

The contents of this book have resulted from a fruitful and highly successful international research cooperation, with which I have been associated since the early 1980s, first as a British Government Assessor of the Heat Pump Programme with India and in later years as a PhD supervisor in the Mexican Programme.

This latter Programme addresses the serious problem of atmospheric pollution in Mexico at its source. The clean combustion of Mexican oils, which are high in sulphur and aspheltenes, requires fundamental research and development studies which extend on the basic chemistry and physics of the fuel to technologies which exploit this information in practical combustion systems. Credit for much of the success of this Programme must be given to Professor F. A. Holland who initiated the work and closely monitored its progress. He also led the complementary programme of research involving heat pumps which is reported herein.

The authors have quite rightly dedicated the book to Dr Pablo Mulás del Pozo PhD (Princeton), Hon. DSc (Salford), Director of Energy Research at the Universidad Nacional Autónoma de México (UNAM) and former Executive Director of the Instituto de Investigaciones Eléctricas (IIE), Mexico, who has, for many years, been a strong advocate of an increased research effort in environmentally clean energy systems. However, credit should also be accorded to two former Directors of the National Chemical Laboratory, Pune, India. First to Dr L. K. Doraiswamy who initiated the cooperation with UK in the late 1970s and then to his successor Dr R. A. Mashelkar, FRS, who continued the cooperation and who subsequently became the Director General of the Council of Scientific and Industrial Research in India.

Professor J. Swithenbank, F Eng

Preface

Research and development work on heat pumps and heat pump assisted purification systems has been carried out between the Instituto de Investigaciones Eléctricas (IIE), Cuernavaca, Mexico, The National Chemical Laboratory (NCL), Pune, India and the University of Salford, UK since 1981, and since 1985 with the Energy Research Centre of the Universidad Nacional Autónoma de México (UNAM).

Following the introduction to this book, Chapters 2 and 3 present the thermodynamic design basis of mechanical vapour compression heat pump systems and absorption heat pump systems respectively. Chapters 4 and 5 describe the process design and fabrication details and the operational characteristics of prototype heat pump assisted pilot purification plants which have been developed and operated in Mexico since the early 1980s. Chapter 6 discusses the economics of heat pump systems and Chapter 7 some alternative purification methods.

The prime purpose of this book is to make widely available the extensive learning experience of the authors in the hope that others can avoid their mistakes in the overall development of the technology.

The authors feel that the current level of development that they have been able to achieve, particularly in absorption heat pump systems, is only because they have had the privilege of being able to stand on the shoulders of giants such as the late Professor Georg Alefeld of the Technical University, Munich, Germany, who so readily gave his advice, and the benefits of his vast experience in heat pump technology on his visits to IIE and UNAM in Mexico, to NCL in India and to the University of Salford in the UK.

However the authors feel that the development of heat pump systems and, in particular, absorption and hybrid heat pump systems is still in its infancy, even with their potential to be highly economic in the immediate future.

Most of the world's population is without sufficiently pure drinking water and this creates serious health problems which have an adverse effect on the economic development of most of the developing countries in the world. Although all the experimental work on heat pump assisted purification systems has been done in Mexico, the results of the work on the experimental pilot plant units are applicable to work in any country in the world, particularly to other developing countries.

This international cooperation in heat pump technology which has developed so profitably in the last two decades would not have been possible without the total support and encouragement of Dr Mulás for each one of us.

The authors also owe a debt of gratitude to many people too numerous to list and also to the British Council, the Overseas Development Administration of the British Government and the Consejo Nacional de Ciencia y Tecnología (CONACyT) for their major financial support.

The authors hope that this financial support so generously given will have made some beneficial contribution to the international community.

1 Introduction

1.1 Nomenclature

(COP)	coefficient of performance [dimensionless]
$(COP)_A$	actual coefficient of performance [dimensionless]
$(COP)_C$	Carnot coefficient of performance [dimensionless]
$(COP)_{CL}$	coefficient of performance for cooling [dimensionless]
$(COP)_H$	coefficient of performance for heating [dimensionless]
$(COP)_{CH}$	Carnot coefficient of performance for an absorption heat pump [dimensionless]
(PER)	primary energy ratio [dimensionless]
Q	heat flow rate [kW] or heat quantity [kWh]
T	temperature [°C or K]
W	rate of work [kW] or quantity of work [kWh]
η	efficiency of conversion [dimensionless]

Subscripts

A	actual
$\bar{A}$	absolute ambient
AB	absorber
CO	condenser
EV	evaporator
GE	generator
H	high
L	low

1.2 Current Problems and Possible Solutions

The scarcity of pure and drinkable water is a major problem throughout Mexico. In addition, the widespread discharge of effluents from industrial processes into lakes and rivers is having an increasingly adverse effect on the environment, as well as posing a health hazard to a significant proportion of the population.

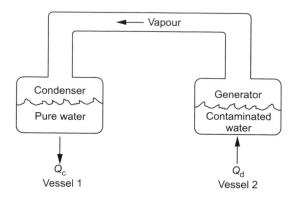

Figure 1.1 Schematic diagram of a heat energy assisted effluent purification system.

The development of both small scale and large scale water purification and effluent concentration systems could make a major contribution to one of Mexico's most pressing problems.

Mexico is a country rich in energy resources, such as oil and gas. Additionally, the large amount of geothermal and solar heat, together with the vast quantities of waste heat discharged from relatively inefficient industrial processes, constitute a huge untapped national resource of low-grade heat energy.

This kind of widely available relatively low-temperature heat could be used to purify water and to concentrate liquid effluents. Consider the two vessels shown in Figure 1.1. Assume that vessel 1 contains pure water and vessel 2 contains water with dissolved solids. If both vessels are at the same temperature, the pure water in vessel 1 has a higher vapour pressure than the impure water in vessel 2. Therefore if both vessels are heated to the same temperature, pure water vapour will flow from vessel 1 to the impure water in vessel 2. However, if the impure water in vessel 2 is heated to a temperature T_2, which is just sufficient to raise its vapour pressure to a higher level than for the pure water in vessel 1 at a temperature T_1, then water will evaporate from the impure water in vessel 2 and flow into vessel 1. This is the principle of heat energy assisted effluent purification.

The temperature difference $(T_2 - T_1)$, which is required to make the effluent purification process work, can be effectively and economically created using a conventional heat pump.

1.3   Heat Pump Systems

A heat pump is a device for raising the temperature of low-grade heat. It can be thought of as a heat engine operating in reverse. Figure 1.2 shows a schematic comparison of a simple heat engine and a simple heat pump operating between two temperature levels T_H and T_L. A heat engine operates

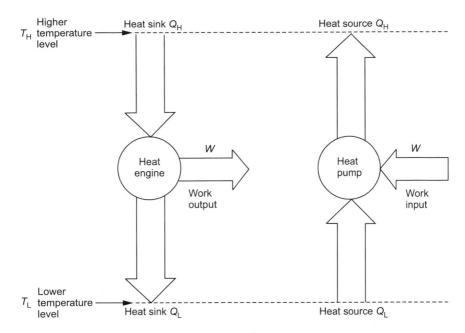

Figure 1.2 Schematic comparison between a simple heat engine and a simple heat pump.

by converting part of a quantity of heat Q_H at a temperature T_H into a quantity of work W whilst delivering the remaining quantity of heat Q_L to a heat sink at a lower temperature T_L.

These quantities are related by the energy balance equation

$$Q_H = Q_L + W \qquad (1.1)$$

and the efficiency of conversion into work is given by the equation

$$\eta = \frac{W}{Q_H} \qquad (1.2)$$

A heat pump operates by taking a quantity of heat Q_L at a temperature T_L, adding a quantity of work W and delivering an increased quantity of heat Q_H to a heat sink at a higher temperature T_H. These quantities are also related by the energy balance equation (1.1). However, the efficiency of a heat pump is measured by the quantity of heat delivered per unit quantity of high-grade energy input. This is known as the coefficient of performance for heating.

$$(COP)_H = \frac{Q_H}{W} \qquad (1.3)$$

A refrigerator or cooler is also a heat pump, but in this case the efficiency is measured by the quantity of heat extracted from a heat source per unit of high-grade energy input. This is known as the coefficient of performance for cooling.

$$(COP)_{CL} = \frac{Q_L}{W} \tag{1.4}$$

A combination of equations (1.1), (1.3) and (1.4) shows that the two coefficients of performance are related by the equation

$$(COP)_H = (COP)_{CL} + 1 \tag{1.5}$$

Both refrigerators and heat pumps produce simultaneous cooling and heating. The difference between them is that refrigerators are specifically designed to produce cooling whilst heat pumps are specifically designed to produce heating. A small domestic refrigerator is designed to extract heat from stored food in order to maintain it at the required low temperature. However, in the process of doing so, the refrigerator delivers heat to the room in which it is situated. In many large-scale commercial and industrial refrigeration systems, the heat delivered is used to provide hot water for heating buildings or for use in industrial processes.

Although there are many different types of heat pump, by far the most common is the mechanical vapour compression heat pump which uses a compressor driven by an electric motor. Currently, the only serious rivals to the mechanical vapour compression heat pump are the heat driven absorption heat pump and the reversed absorption heat pump or heat transformer. These three heat pump systems are the only ones to be considered in this book, since most of the other systems have little or no commercial relevance at the present time.

1.3.1   *Mechanical vapour compression heat pumps*

A schematic line diagram of a conventional mechanical vapour compression heat pump is shown in Figure 1.3. It consists of two heat exchangers, a compressor, an expansion valve and a refrigerant or working fluid. In the evaporator heat exchanger, the working fluid evaporates at an absolute temperature T_{EV} whilst extracting an amount of heat Q_{EV} from the heat source which may be in the gaseous, liquid or solid state. The working fluid is then compressed and, as it condenses, gives up an amount of latent heat Q_{CO} at a higher absolute temperature T_{CO} in the condenser heat exchanger. The condensed working fluid is then expanded through the expansion valve and is returned to the evaporator to complete the cycle. The cycle involves two temperature and two pressure levels which are conventionally illustrated on a plot of pressure against temperature as shown in Figure 1.4. However, the

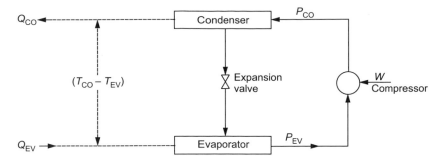

Figure 1.3 Schematic diagram of a mechanical vapour compression heat pump.

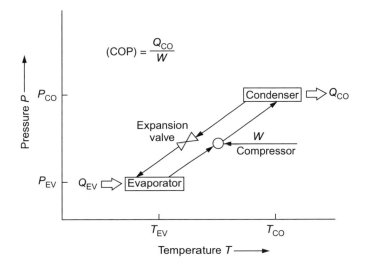

Figure 1.4 Schematic diagram of a mechanical vapour compression heat pump in a plot of pressure P against temperature T.

importance of temperature is more clearly illustrated on a plot of temperature against pressure as shown in Figure 1.5.

An energy balance on the system gives

$$Q_{CO} = Q_{EV} + W \qquad (1.6)$$

where W is the input of mechanical energy from the compressor. The coefficient of performance (COP) of a mechanical vapour compression heat pump is the ratio of heat given out in the condenser Q_{CO} to the work put in by the compressor W.

$$(COP) = \frac{Q_{CO}}{W} \qquad (1.7)$$

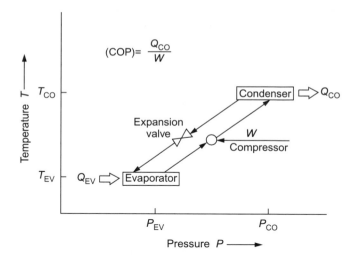

Figure 1.5 Schematic diagram of a mechanical vapour compression heat pump in a plot of temperature T against pressure P.

Equation (1.7) can also be written as

$$(COP) = \frac{Q_{CO}}{Q_{CO} - Q_{EV}} \tag{1.8}$$

For a thermodynamically reversible system in which there is no increase in entropy, an ideal theoretical Carnot coefficient of performance can be defined as

$$(COP)_C = \frac{T_{CO}}{T_{CO} - T_{EV}} \tag{1.9}$$

$(COP)_C$ is the ratio of the condensing temperature of the working fluid to the gross temperature lift $(T_{CO} - T_{EV})$ produced by the heat pump. The net or effective temperature lift $(T_D - T_S)$, which is the difference between the delivery and source temperatures is, of course, less than $(T_{CO} - T_{EV})$ by the sum of the temperature difference driving forces in the condenser and evaporator heat exchangers.

The coefficient of performance (COP) of a mechanical vapour compression refrigerator is the ratio of the heat absorbed in the evaporator Q_{EV} to the work put in by the compressor W.

$$(COP) = \frac{Q_{EV}}{W} \tag{1.10}$$

The corresponding ideal Carnot coefficient of performance for a mechanical vapour compression refrigerator can be defined as

$$(\text{COP})_C = \frac{T_{EV}}{T_{CO} - T_{EV}} \qquad (1.11)$$

The Carnot coefficient of performance $(\text{COP})_C$ depends only on the temperatures in the system and is independent of the working fluid used. It can never be achieved in practice.

1.3.2   Heat driven absorption heat pumps

Absorption heat pumps are devices for raising the temperature of low-grade heat to a more useful level using a smaller amount of relatively high-grade heat energy. The higher the absolute temperature T of a quantity of heat Q, the lower its entropy Q/T and the greater its quality as measured by its exergy or work equivalent W given by the equation

$$W = Q\left(1 - \frac{T_{\bar{A}}}{T}\right) \qquad (1.12)$$

with reference to an absolute ambient temperature $T_{\bar{A}}$.

In the heat driven absorption system shown in Figure 1.6, the condensation, expansion and evaporation of the working fluid are the same as in the mechanical vapour compression system. In the absorption cycle, however, the compressor is replaced by a secondary circuit in which a liquid absorbent is circulated by a pump. The evaporated working fluid is absorbed by the circulating liquid and the pressure increased by the pump prior to entering the generator. An amount of heat Q_{GE} is added at a temperature T_{GE} in the

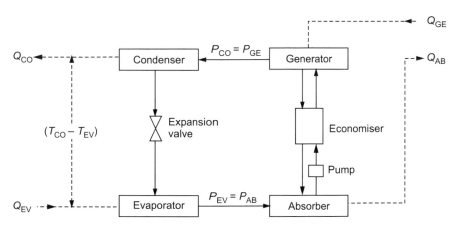

Figure 1.6 Schematic diagram of a heat driven absorption heat pump.

generator to produce the high-pressure working fluid vapour required to be fed to the condenser. The mechanical energy required to pump the liquid is usually negligible compared with the input of high-grade heat energy Q_{GE}. An actual coefficient of performance for an absorption heat pump can be defined as

$$(COP)_A = \frac{Q_{CO} + Q_{AB}}{Q_{GE}} \tag{1.13}$$

where Q_{CO} and Q_{AB} are the amounts of heat delivered at the condenser and the absorber, respectively.

Mechanical vapour compression heat pumps have higher coefficients of performance than heat driven absorption heat pumps. However when comparing heat pump systems driven by different energy sources, it is more appropriate to use the primary energy ratio defined as

$$(PER) = \frac{\text{Useful heat delivered}}{\text{Primary energy input}} \tag{1.14}$$

Based on a comparison of PERs rather than COPs, heat driven absorption heat pumps are often as energy efficient as mechanical vapour compression heat pumps and sometimes more so. They also have the advantage of operating with environmentally friendly working fluids compared with the chlorofluorocarbons traditionally used in mechanical vapour compression heat pumps.

Equation (1.15) is the ideal Carnot coefficient of performance for a thermodynamically reversible absorption heat pump cycle.

$$(COP)_{CH} = \left(1 - \frac{T_{EV}}{T_{GE}}\right)\left(\frac{T_{CO}}{T_{CO} - T_{EV}}\right) \tag{1.15}$$

A comparison of equations (1.9) and (1.15) shows that the Carnot coefficient of performance of a thermodynamically reversible absorption heat pump is less than the Carnot coefficient of performance of a thermodynamically reversible mechanical vapour compression system by a factor which is the Carnot cycle efficiency of a thermodynamically reversible heat engine taking in heat at an absolute temperature T_{GE} and rejecting it at an absolute temperature T_{EV}. Thus, thermodynamically, an absorption heat pump is equivalent to a heat engine driving a compressor driven vapour compression heat pump. Equation (1.15) shows that the theoretically ideal Carnot coefficient of performance $(COP)_{CH}$ is independent of the working pair and is dependent only on the temperatures T_{EV}, $T_{CO} = T_{AB}$ and T_{GE}.

In practice, the Carnot coefficient of performance can never be achieved because of the unavoidable irreversibilities that arise from the operation of a practical system. For example, the net temperature lift for a heat pump is less than the gross temperature lift $(T_{CO} - T_{EV})$ because of the required temperature difference driving forces in the heat exchangers.

1.3.3 Heat driven heat transformers

A heat driven heat transformer, also known as a reversed absorption heat pump or temperature amplifier, is shown schematically in Figure 1.7. It differs from a conventional absorption heat pump in that the evaporator operates at a higher temperature than the condenser. It also employs two pumps instead of one.

The driving force for a heat transformer consists of heat inputs to the generator and the evaporator. The heat transformer then delivers part of the heat at a higher temperature from the absorber and the rest at a lower temperature from the condenser. However, the unique ability of the heat transformer to deliver heat at a higher temperature than the input heat, depends upon the availability of a lower temperature heat sink. The lower the temperature of the heat discharged at the condenser, the higher the temperature of the heat delivered at the absorber.

Quantities of heat Q_{GE} and Q_{EV} are supplied to the generator and evaporator respectively at the evaporator temperature T_{EV}. A quantity of heat Q_{AB} is given out at the absorber at the higher absorber temperature T_{AB} together with a quantity of heat Q_{CO} at the condenser at the lower condenser temperature T_{CO}. The coefficient of performance of a heat transformer is the heat given out at the absorber Q_{AB} divided by the total amount of heat supplied to the generator and the evaporator $(Q_{GE} + Q_{EV})$.

$$(\text{COP}) = \frac{Q_{AB}}{Q_{GE} + Q_{EV}} \tag{1.16}$$

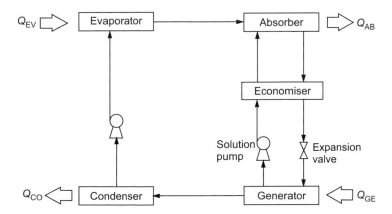

Figure 1.7 Schematic diagram of a heat driven heat transformer.

Although the coefficient of performance of a heat transformer is low, in the region of 0.4, it has the unique ability to raise the temperature of otherwise waste heat to a more useful level with a negligible input of additional energy required for the pumps.

Heat transformers are being used increasingly on an industrial scale to provide environmentally clean steam from waste heat. Heat driven absorption heat pumps and heat transformers have some major advantages over mechanical vapour compression heat pumps:

- They use very small amounts of mechanical energy.
- Working fluids which have no adverse effect on the environment can be used.
- They do not consist of any high technology items, such as compressors, which are not readily available in many developing countries.
- They can readily be designed by any competent engineer provided that he or she has the necessary thermodynamic data.

1.3.4   *Comparison of energy inputs and outputs for the common types of heat pumps*

The primary energy ratio (PER) is a more fundamental measure of the energy efficiency of a heat pump than the coefficient of performance (COP). Figure 1.8 is a schematic comparison of the energy inputs and outputs for the three main types of heat pumps.

The primary energy used to generate the electricity employed to drive the compressor in a mechanical vapour compression heat pump could be oil, coal or natural gas. Alternatively, an oil or gas driven engine could be used to drive the compressor. Although a heat driven absorption heat pump has a

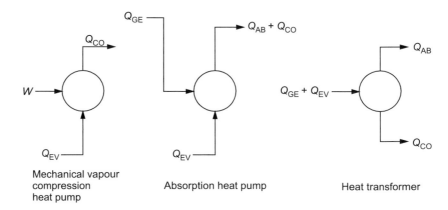

Figure 1.8 Schematic comparison of the energy inputs and outputs for the three main types of heat pumps.

significantly lower coefficient of performance than a mechanical vapour compression heat pump, it can have a primary energy ratio (PER) equal to, or even greater than, a mechanical vapour compression heat pump driven by electricity.

Consider an electrically driven mechanical vapour compression heat pump with a coefficient of performance (COP) = 5.0. If the electricity was generated in an oil burning power generating station with an overall efficiency $\eta_{OV} = 0.3$ or 30%, the primary energy ratio of the heat pump, based on the primary energy of oil (PER) = $0.3 \times 5.0 = 1.5$.

An absorption heat pump driven directly by an oil burner could have a comparable primary energy ratio. However, the primary energy ratio of a mechanical vapour compression heat pump can be increased by using an engine to drive the compressor.

1.4 Heat Pump Assisted Purification Systems

Since the early 1980s, extensive experimental work has been carried out by researchers at the Instituto de Investigaciones Eléctricas (IIE), Cuernavaca, Morelos, Mexico, on pilot plants for producing pure water from geothermal brine using electrically driven mechanical vapour compression heat pumps. A simplified schematic diagram of the systems used is shown in Figure 1.9 [Frias, 1991; Frias *et al.*, 1991; Siqueiros *et al.*, 1995].

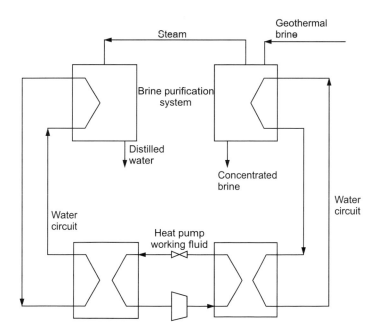

Figure 1.9 Simplified schematic diagram of a mechanical vapour compression heat pump assisted purification system.

These IIE pilot plants have worked extremely well. The quality of the distilled water obtained from the geothermal brine, with a typical salt concentration of 2.3%, was similar to commercially available distilled water with respect to chlorides and silica. An economic assessment for the production of distilled water from geothermal brine in Mexico has produced a favourable result with payback periods of between six months and three years [Siqueiros *et al.*, 1992].

Because of the scarcity of electricity in parts of Mexico, the difficulty in obtaining the most suitable compressors and the existence of large unused resources of low-grade heat throughout Mexico, it was decided to design a heat driven absorption heat pump to replace the mechanical vapour

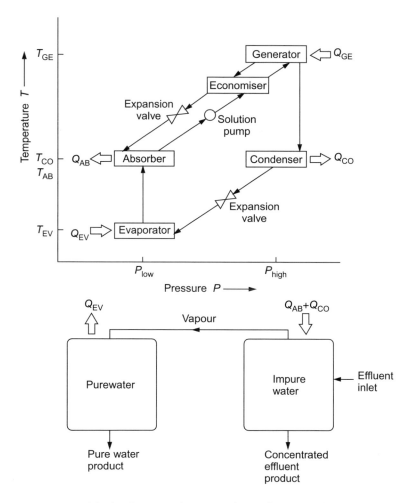

Figure 1.10 Simplified schematic diagram of an absorption heat pump assisted purification system.

compression heat pumps in the IIE heat pump assisted purification systems. A simplified schematic diagram of an absorption heat pump assisted purification system is shown in Figure 1.10.

1.5   Water as a Universal Resource

Water covers 70% of the Earth's surface in the form of oceans. The water from the oceans contains about 3.5% by weight of dissolved materials but in some cases this can be as high as 42 000 mg kg^{-1} (4.2%) as in the Persian Gulf. About 50% of the Sun's energy falling on the ocean causes the water to evaporate. Thus, the vapours form clouds that precipitate pure water as rain.

Most rain falls on the ocean while a small portion falls on the land, producing the hydrological cycle. In the hydrological cycle, land rainfall returns to the sea through rivers or percolates into the ground and flows back to the sea, or is evaporated. This is a natural almost closed distillation cycle which only has additions from volcanic activity and from space. Small amounts of water can also be lost due to chemical reactions and dissociated hydrogen is lost to space.

Withdrawal use is the quantity of water removed from the ground or diverted from a body of surface water. Consumptive use is the portion of the water that is discharged to the atmosphere or incorporated in food products or in industrial processes or in growing vegetation. The demand for water is increasing world-wide. For example, Darwish and El-Dessouky [1996] reported that the per capita water consumption in Kuwait increased from 0.057 m^3 d^{-1} in 1958 to 0.475 m^3 d^{-1} in 1994. Global replenishment of accessible fresh water supplies is equivalent to 1800 m^3 person^{-1} y^{-1}, some 50 times the recommended minimum international standard for human needs [Ayoub, 1996]. However, distribution of the rainfall is uneven, and many countries now face, or will be soon experiencing, water supply and quality problems.

The maximum supply of water available daily is commonly called runoff and properly includes both surface and underground flows. Fresh water withdrawal is about 30% of runoff and consumptive use is about 8% of runoff. The consumptive use of water for irrigation is about 55%, but another 15% is allowed for transmission and distribution losses.

The withdrawal and consumption of water for domestic purposes is relatively small compared with industrial and agricultural uses. This can be illustrated by the 0.12 m^3 per capita per day which was the average domestic consumption of water in urban areas of the United States of America in 1976. However, since public water utilities also supply industrial and commercial customers, the average withdrawal by US municipal water systems was 0.56 m^3 per capita per day in 1976 [Roe and Vernick, 1978]. Additions to water resources for the future can be made as follows:

- increase in storage reservoirs;
- injection of used water or flood water into natural aquifers;
- covering reservoirs with films to reduce evaporation;
- rainmaking;
- saline water-conversion;
- waste water renovation and reuse.

Also it is important to improve the efficient use of water supplies as follows:

- multiple use of cooling water;
- use of air cooling instead of water cooling;
- use of cooling towers;
- reclamation of waste waters, both industrial and domestic;
- abatement of pollution by treatment instead of dilution which requires additional fresh water.

1.5.1   Water in industry

Water is an indispensable material for most chemical process operations. The use of water in industry varies widely because of differing conditions of price, availability, and process technology. When a sufficient water supply of suitable quality is available at low cost, industry tends to use maximum volumes. However, when water is scarce and costly, approved processes and careful water management can reduce the water usage to the minimum.

Water must be treated to the industry standards dictated by the end use. About 30% of industrial water is used for cooling, mostly on a once-through

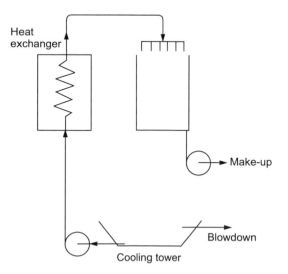

Figure 1.11 Schematic diagram of an open recirculation water cooling system.

basis. Figure 1.11 shows an open recirculation system with a cooling tower or spray pond. This system reduces the withdrawal use of water by over 90% but increases consumptive use by 3–8% because of evaporation losses. A greater reduction in water demand can be achieved using multiple reuse (Figure 1.12). Thus, it is important to establish some water quality requirements for chemical plants, especially where the water is used for drinking. Water must be low in suspended solids and total dissolved solids to avoid clogging and depositions, respectively. These water requirements are important in the case of cooling services.

When the water is high in suspended solids, a reduction is achieved by settling or by using a coagulating additive, such as alum, and then settling. In the case of recirculating systems, the water is treated with corrosion inhibitors such as polyphosphates or chromates. Algaecides and biocides are used to avoid and to control the growth of micro-organisms. Hardness of water may cause scaling in the equipment. For this reason, the process water requirements are better defined than the potable water standard. For example, in the case of boiler feedwater, it must have less than $1\,\mathrm{mg\,kg^{-1}}$ of dissolved solids. Thus, the required quality may be met by the general available plant water or it must be provided by treatment.

Table 1.1 lists the most common methods used for industrial water treatment to remove a range of contaminants. Modern treatment methods

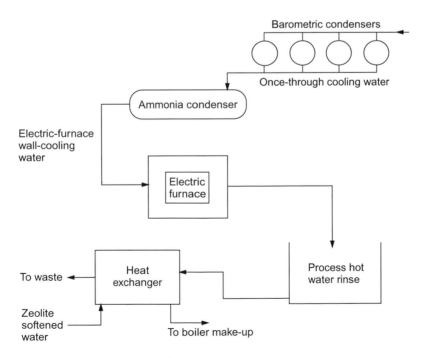

Figure 1.12 Schematic diagram of a multiple reuse water system.

Table 1.1 Most common methods used for industrial water treatment [modified from *UK Environment Protection Act 1990* (1997)]

Technique	Purpose
Biological treatment	
Anaerobic digestion	Oxidises organic matter in closed vessels in the absence of air
Aerobic biological process	Oxidises organic matter in open vessels in the presence of air
Physical treatment	
Adsorption	
Zeolites	Removes ammonia or toxic metals such as cadmium from wastewaters
Carbon activated	Removes organic compounds
Electrodialysis	Partially removes iron; can reduce $10\,000$ mg kg^{-1} brackish water to 500 mg kg^{-1} or less
Evaporation using steam heat	Produces very pure water 10 mg kg^{-1} or less total solids
Filtering	Removes suspended solids
Pre-sedimentation	Removes suspended solids and reduces turbidity and colour
Reverse osmosis (RO)	Partially removes iron; can reduce the content of seawater and brackish water to 500 mg kg^{-1} or less
Settling	Removes oils and greases by flotation in the quiescent conditions
Chemical treatment	
Advanced oxidation process	Removes organic compounds from wastes containing between 3 and 20% solids using short retention times
Chlorine dosing	Prevents algae and slime growth (additions of chlorine 5–6 mg kg^{-1} or continuous feed to maintain 0.2–0.3 mg kg^{-1} residual free Cl_2)
Coagulation	Removes suspended solids, some dissolved organic compounds and micro-organisms
Electrodeposition	Removes plateable metals, i.e. zinc, copper, cadmium from solutions
High density sludge process	Removes heavy metals such as cadmium from acidic waste-streams with a pH of 4 and below
Ion exchange: two stages or mixed bed	Removes both positive and negative ions (cations and anions) to provide very pure water
Lime dosing	Removes phosphates, sulphates, precipitating insoluble hydroxide products

include clarification, coagulation, filtration, membrane separation, ion exchange, ultrafiltration and chemical treatment [DeSilva, 1996]. The source water strongly affects the choice of water treatment scheme. When considering the options, the capital costs, maintenance and operating costs, and the need for any necessary pre-treatment systems must be assessed.

1.5.2   Water pollution control

One of the unfortunate characteristics of any modern society is the rapidly increasing discharge of toxic waste streams to the environment. The treatment of any effluent stream normally relies upon a series of unit operations and these operations fall into three broad categories: (a) biological treatment: aerobic, anaerobic; (b) physical treatment: filtration, flotation, settlement; (c) chemical treatment: precipitation, pH correction, adsorption [Glancy, 1997].

Pollutants include dredge spoil, solid waste, incinerated residue, sewage, garbage, sewage sludge, munitions, chemical wastes, biological materials, heat, wrecked or discarded equipment, rock, sand, cellar dirt and industrial, municipal, and agricultural waste discharged into water.

The discharge of industrial wastewater is regulated for applicable pollutant parameters, such as suspended solids, biodegradable oxygen demand (BOD), chemical oxygen demand (COD), colour, pH, oil and grease, metals, ammonia and phenol. In order to limit these pollutants, most industrial facilities must treat their wastewater effluent by either physical, chemical or biological methods.

Physical treatment includes screening, settling, flotation, equalisation, centrifuging, filtration and carbon adsorption.

Chemical treatment comprises coagulation or neutralisation of acids with soda ash, caustic soda or lime. Alkali wastes are treated with sulphuric acid or inexpensive waste acids for neutralisation. Chemical oxidation is effective for certain wastes.

Biological treatment is accomplished by the action of two types of micro-organisms. The aerobic micro-organisms act in the presence of oxygen. The anaerobic micro-organisms act in the absence of oxygen and are employed in processes for the digestion of the sludge produced by biological wastewater treatment. Most organic wastes can be treated by biological processes. The more common wastewater treatment techniques include the activated sludge process and aerated lagoons which all employ aerobic micro-organisms to degrade the organic waste material.

1.6   References

Ayoub, J. R. (1996) Water requirements and remote arid areas: The need for small-scale desalination, *Desalination*, **107**(2), 131–47.

Darwish, M. A. and El-Dessouky, H. (1996) The heat recovery thermal vapour–compression desalting system: A comparison with other thermal desalination processes, *Applied Thermal Engineering*, **16**(6), 523–37.

DeSilva, F. (1996) Tips for process water purification, *Chemical Engineering*, **103**(8), 9.

Frias, J. L. (1991) An experimental study of a heat pump assisted purification system for geothermal brine, MSc Thesis, University of Salford, UK.

Frias, J. L., Siqueiros, J., Fernandez, H., Garcia, A. and Holland, F. A. (1991) Developments in geothermal energy in Mexico - Part 36: The commissioning of a heat pump assisted brine purification system, *Heat Recovery Systems*, **11**(4), 297–310.

Glancy, V. (1997) Effluent treatment and disposal, *Chemical Engineer*, **633**, 29–35.

Roe, K. A. and Vernick, A. S. (1978) Water, *Mark's Standard Handbook for Mechanical Engineers*, 3rd edn (ed. T. Baumeister), McGraw-Hill, New York, USA, pp. 187–97.

Siqueiros, J., Heard, C. and Holland, F. A. (1995) The commissioning of an integrated heat-pump assisted geothermal brine purification system, *Heat Recovery Systems*, **15**(7), 655–64.

Siqueiros, J., Fernandez, H., Heard C. and Barragan D. (1992) Desarrollo e implantacion de tecnologia de bombas de calor, *Final Report: INFORME IIE/FE/11/2963/F*, Cuernavaca, Mexico.

UK Environmental Protection Act 1990 (1997), *Technical Guidance Note (Abatement) A4: Effluent Treatment Techniques*, The Stationery Office, London, UK, p. 86.

2 Design basis for mechanical vapour compression heat pump systems

2.1 Nomenclature

(COP)	coefficient of performance [dimensionless]
$(COP)_A$	actual coefficient of performance [dimensionless]
$(COP)_C$	Carnot coefficient of performance [dimensionless]
$(COP)_R$	Rankine coefficient of performance [dimensionless]
(CR)	compression ratio [dimensionless]
F	volumetric flow rate [$m^3\ s^{-1}$]
H	enthalpy per unit mass [$kJ\ kg^{-1}$]
(HPE)	heat pump effectiveness [dimensionless]
P	absolute pressure [kPa]
(PER)	primary energy ratio [dimensionless]
Q	heat flow rate [kW] or heat quantity [kWh]
T	temperature [°C or K]
(VPH)	vapour volume per unit latent heat of condensation [$m^3\ MJ^{-1}$]
W	rate of work [kW] or quantity of work [kWh]
η_{ov}	overall efficiency [dimensionless]
η_{vol}	volumetric efficiency [dimensionless]

Subscripts

A	actual
C	critical
CO	condenser
D	delivered
EV	evaporator
R	Rankine
S	source

2.2 Introduction

There is a considerable amount of thermodynamic data available in the literature for the process design of mechanical vapour compression heat pump systems [Holland *et al.*, 1982]. Unfortunately, most of these data are for chlorofluorocarbons (CFCs) and hydrochlorofluorocarbons (HCFCs) which are being phased out in many countries because of their adverse effect on the environment.

Any new materials which are being considered to replace CFCs and HCFCs as refrigerants or working fluids in heat pump systems must have a low ozone depleting potential (ODP) and a low global warming potential (GWP).

Performance data will be required for any new working fluids to replace the widely available data on the most commonly used working fluids to date, such as R12 (dichlorodifluoromethane CCl_2F_2) and R114 (dichlorotetra-fluoroethane $CClF_2\ CClF_2$), which are used up to condensing temperatures of about 70°C and 120°C respectively. However, not only are the replacement working fluids more expensive, but there are additional complications such as the need for specially developed lubricants and changes in the physical components of the heat pump systems.

HFE 134 ($C_2\ H_2F_4O$) is considered to be the best alternative to replace R12 in heat pump and refrigeration applications [Devotta, 1995]. HFE 134 has an ozone depleting potential (ODP) of zero compared with 1.0 for R12. HFE 134 (CHF_2OCHF_2) and HFC 143 ($CHClF_2$) are potential replacements for R114. Both have ODPs of zero compared with 0.8 for R114.

For high-temperature heat pumps with condensing temperatures greater than $T_{CO} = 150$°C, the best working fluid is water R718 (H_2O) with its entirely environmentally friendly characteristics.

2.3   The Basic Design Parameters

Figure 2.1 is a schematic diagram of a mechanical vapour compression heat pump in the context of a plot of pressure against temperature.

There are two pressure levels and two temperature levels in a mechanical vapour compression heat pump system. The working fluid or refrigerant is condensed in the condenser producing a quantity of heat Q_{CO} at an absolute temperature T_{CO} with an entropy value of Q_{CO}/T_{CO}. After passing through the expansion valve and having its pressure reduced from P_{CO} to P_{EV}, the working fluid is evaporated in the evaporator, absorbing a quantity of heat Q_{EV} at an absolute temperature T_{EV} with an entropy value of Q_{EV}/T_{EV}.

For thermodynamically reversible processes in the condenser and the evaporator, the reduction in entropy in the condenser would equal the gain in entropy in the evaporator so that

$$\frac{Q_{CO}}{T_{CO}} = \frac{Q_{EV}}{T_{EV}} \tag{2.1}$$

An energy balance on the compressor-driven system shown in Figure 2.1 gives:

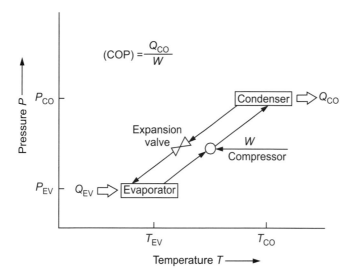

Figure 2.1 Schematic diagram of a mechanical vapour compression heat pump in a plot of pressure *P* against temperature *T*.

$$Q_{CO} = Q_{EV} + W \tag{2.2}$$

where *W* is the input of mechanical energy to the compressor required to produce the compression ratio

$$(CR) = \frac{P_{CO}}{P_{EV}} \tag{2.3}$$

The coefficient of performance (COP) of a mechanical vapour compression heat pump is the ratio

$$(COP) = \frac{Q_{CO}}{W} \tag{2.4}$$

Equation (2.4) can also be written in terms of equation (2.2) as

$$(COP) = \frac{Q_{CO}}{(Q_{CO} - Q_{EV})} \tag{2.5}$$

For a thermodynamically reversible system, in which there is no increase in entropy, an ideal Carnot coefficient of performance $(COP)_C$ can be obtained by substituting equation (2.1) into equation (2.5) to give

$$(COP)_C = \frac{T_{CO}}{(T_{CO} - T_{EV})} \tag{2.6}$$

which is ratio of the condensing temperature T_{CO} of the working fluid to the gross temperature lift $(T_{CO} - T_{EV})$ produced by the heat pump.

The Carnot coefficient of performance $(COP)_C$ depends only on the temperatures in the system and is independent of the working fluid or refrigerant used. It can never be achieved in practice. In practice, the operation of a mechanical vapour compression heat pump depends on the thermodynamic properties of the working fluid used. Its operation approximates more closely to the Rankine heat pump cycle than to the theoretical Carnot cycle. The ideal Rankine heat pump cycle, which is the reverse of the Rankine power cycle, can be illustrated on a plot of pressure P against enthalpy per unit mass H as shown in Figure 2.2. A theoretically ideal Rankine coefficient of performance $(COP)_R$ can be written with reference to Figure 2.2 as

$$(COP)_R = \frac{H_{D1} - H_{D3}}{H_{D1} - H_{S2}} \tag{2.7}$$

The critical temperature T_C of the working fluid provides the upper limit at which a condensing vapour heat pump can deliver heat energy. The commonly used working fluids R12 and R114 have critical temperatures $T_C = 112.0°C$ and $T_C = 145.7°C$ and critical pressures $P_C = 41.15$ bar and $P_C = 32.63$ bar respectively. R718 (water) has a critical temperature $T_C = 373.0°C$ and a critical pressure $P_C = 221.2$ bar. More detailed thermodynamic data for R718, R12 and R114 are given in Appendices 1, 2 and 3 respectively.

Ammonia (R717) is a commonly used working fluid in refrigeration systems. It has a critical temperature $T_C = 132.2°C$ and a very high critical

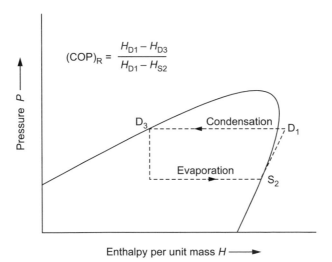

Figure 2.2   The ideal Rankine heat pump cycle in a plot of pressure P against enthalpy per unit mass H.

pressure $P_C = 114.3$ bar. More detailed thermodynamic data for R717 are given in Appendix 4.

There are four important parameters to consider when selecting a working fluid for a vapour compression heat pump system:

- the temperature of condensation T_{CO};
- the compression ratio $(CR) = P_{CO}/P_{EV}$;
- the gross temperature lift $(T_{CO} - T_{EV})$;
- the theoretical Rankine coefficient of performance $(COP)_R$.

These parameters are not independent and when two are specified, the other two are determined automatically for a given working fluid. The first parameter T_{CO} is largely determined by the required delivery temperature T_D since the temperature difference driving force in the condenser is $(T_{CO} - T_D)$.

The second parameter $(CR) = P_{CO}/P_{EV}$ is often determined by the characteristics of an available compressor. Therefore, in practice, one may have little control over the values for the gross temperature lift $(T_{CO} - T_{EV})$ and the theoretical Rankine coefficient of performance $(COP)_R$ since these are a function of the thermodynamic properties of the chosen working fluid.

The net temperature lift between the delivered and the heat source temperature $(T_D - T_S)$ is, of course, less than the gross temperature lift $(T_{CO} - T_{EV})$ by the sum of the temperature difference driving forces in the condenser and evaporator heat exchangers. The actual coefficient of performance $(COP)_A$ is also less than the theoretical Rankine coefficient of performance $(COP)_R$ because of the thermodynamic irreversibilities in the system and the difference in both cases is a function of the process design and the equipment.

A heat pump effectiveness compared with the ideal theoretical Rankine cycle can be defined as

$$(HPE)_R = \frac{(COP)_A}{(COP)_R} \tag{2.8}$$

The $(HPE)_R$ is effectively the ratio of work required in an ideal Rankine cycle to that in an actual cycle in order to transfer a given amount of heat to the condenser. It is possible to exceed $(HPE)_R$ values of 0.8 in a well-designed heat pump system.

Figure 2.3 is a plot of the theoretical Rankine coefficient of performance $(COP)_R$ against the condensing temperature T_{CO} for various values of the compression ratio (CR) and the gross temperature lift $(T_{CO} - T_{EV})$ for water (R718). Figure 2.3 illustrates the design constraints for any particular working fluid.

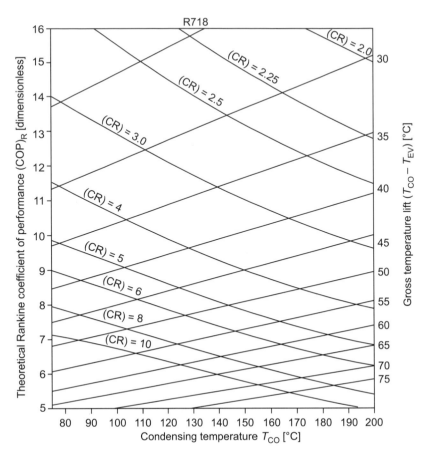

Figure 2.3 Plot of theoretical Rankine coefficient of performance $(COP)_R$ against
condensing temperature T_{CO} for various values of compression ratio (CR)
and gross temperature lift $(T_{CO} - T_{EV})$ for R718.

Plots of $(COP)_R$ against T_{CO} for various values of (CR) and $(T_{CO} - T_{EV})$
for R12, R114 and R717 are given in Appendices 2, 3 and 4 respectively.

For example, consider an application in which the required condensing
temperature $T_{CO} = 170°C$ and the available compressor has a compression
ratio (CR) = 4. If R718 is employed as a working fluid, then from Table A1.2
the gross temperature lift $(T_{CO} - T_{EV}) \approx 50°C$ and the theoretical Rankine
coefficient of performance $(COP)_R \approx 8.42$ are automatically established by
the constraints of the thermodynamic data.

Figure 2.4, based on a compression ratio (CR) = 4, is a plot of $(T_{CO} - T_{EV})$
against T_{CO} for the working fluid R12, R114 and R718 which have critical
temperatures T_C of 112.0°C, 145.7°C and 373.0°C respectively. This
plot further illustrates the design constraints of any particular working fluid.

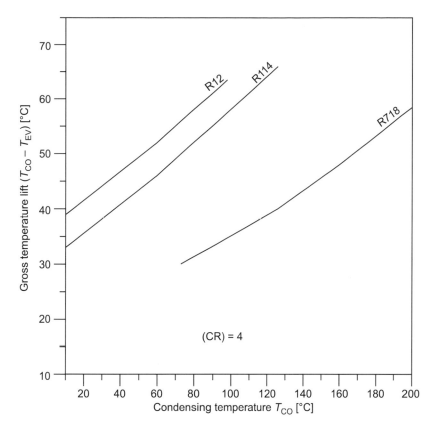

Figure 2.4 Plot of gross temperature lift $(T_{CO} - T_{EV})$ against condensation temperature T_{CO} for R12, R114 and R718 for a compression ratio (CR) = 4.

2.4   Deviations from the Ideal Rankine Cycle

The ideal Rankine cycle is not 'ideal' in the same sense as the Carnot cycle
which represents the performance of a truly reversible heat engine. Because
the Carnot cycle operates infinitely slowly, the net change in entropy per cycle
is zero. The ideal Rankine cycle also operates infinitely slowly, because of the
implied lack of temperature difference driving forces in the evaporator and
condenser heat exchangers. However, there is an increase in entropy in the
Rankine cycle.

The coefficient of performance of an actual mechanical vapour compression heat pump can be written as

$$(COP)_A = \frac{Q_D}{W} \tag{2.9}$$

where Q_D is the quantity or rate of heat delivered to the application and W is the quantity or rate of work supplied by the compressor.

Superheating, subcooling and pressure drops all introduce deviations from the ideal Rankine cycle so that an actual cycle on a pressure against enthalpy plot looks more like Figure 2.5 than the theoretical ideal Rankine cycle shown in Figure 2.2. In Figure 2.5, the effect of various non-idealities has been exaggerated for clarity.

From D′1 to D′2, the superheat is removed in the condenser and, as the heat transfer coefficient for desuperheating is much less than that for condensation, a considerable portion of the condenser surface can be involved in this stage. Provided that the temperature and pressure changes are not too large between D′1 and D′2, the specific volume of the vapour at the means of the temperatures and pressures between these two points may be used to estimate the pressure drop in the desuperheating section. In the absence of this pressure drop, the latent heat would have been delivered at a higher temperature and pressure, thus increasing the (COP) attainable.

From D′2 to D′3, the vapour is condensed and, if excess surface area is available, the condensed liquid is subcooled by the medium to which the heat is being transferred. In practice, condensation takes place immediately upon entry to the condenser so that desuperheating, condensation and subcooling take place throughout. It is usually satisfactory, since pressure drops in the latter two stages are relatively small, to consider that all the pressure drop takes place during superheating and that condensation and subcooling take

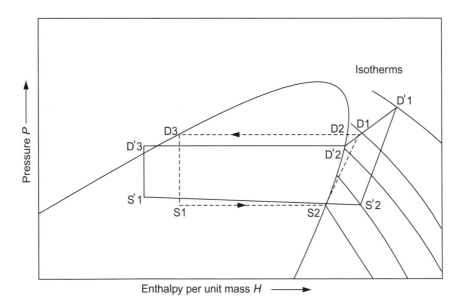

Figure 2.5 Deviations from the ideal Rankine heat pump cycle in a plot of pressure P against enthalpy per unit mass H.

place at the pressure of the condenser outlet, i.e. that D'3 is in horizontal alignment with D3. The location of D'3 can be established by measurement of the temperature and pressure at the outlet of the condenser.

From D'3 to S'1, the expansion will take place almost isenthalpically between the condenser and the evaporator. A small pressure drop is needed to progress the wet vapour along the evaporator and, if excess surface area is available, to permit the progress of the superheated vapour to the inlet of the compressor at S'2. It is usually satisfactory to consider that the line between S'1 and S'2 is horizontal at the outlet pressure of the evaporator as both evaporation and superheating take place throughout the evaporator.

From S'2 to D'1, the vapour is compressed. The energy required in a real compressor will be greater than that in an isentropic compressor for two reasons. Firstly, compression takes place in a finite time, and therefore irreversibly, the excess energy appearing as heat when the turbulence dies away. Secondly, the volumetric efficiency is less than unity which effectively means that part of the vapour is repeatedly compressed and hence absorbs energy on each cycle, only some of which is recovered on the expansion stroke. The exact amount of excess entropy, and hence enthalpy, due to these causes depends on the compressor design and on the thermodynamic properties of the working fluid but the location of the point D'1 can be established by measurement of the temperature and pressure at the outlet of the compressor. The value of Q_D in equation (2.9) is the difference in enthalpy between points D'1 and D'3 while the value of W in equation (2.9) is the measured shaft work into the compressor.

2.5 The Choice of a Working Fluid

The selection of a suitable working fluid is a critical step in the design of a heat pump system. An ideal working fluid should be:

- thermodynamically and physically favourable;
- thermally and chemically stable;
- safe (non-flammable, non-toxic, non-explosive);
- readily available and not too expensive;
- compatible with the materials of construction and lubricant used in the compressor.

The selection of a working fluid for a particular application is always a compromise between the various factors.

The critical temperature T_C of the working fluid provides the upper limit at which a condensing vapour heat pump can deliver heat energy. The working fluid should be condensed at a temperature sufficiently below the critical temperature to provide an adequate amount of latent heat per unit mass. As the condensing temperature approaches the critical temperature of the working fluid, the latent heat of vaporisation decreases rapidly. Consequently, the amount of working fluid required to be circulated increases. This implies an increase in the work of compression per unit of heat transferred, so that the coefficient of performance decreases.

The condensing pressure of the working fluid should also be significantly below its critical pressure. Moderate condensing pressures allow the use of lightweight materials in construction, thereby reducing the size, weight and cost of the equipment. However, the evaporating pressure should be above atmospheric to avoid air leaks into the system. Therefore, the working fluid should have a reasonably high evaporating pressure P_{EV} and a relatively low condensing pressure P_{CO}, implying a low compression ratio $(CR) = P_{CO}/P_{EV}$. Low compression ratios result in a low power consumption for the compressor and a high volumetric efficiency.

The latent heat of vaporisation of the working fluid should be as large as possible at the normal operating temperature of the system so that the maximum amount of heat can be transferred for a given circulation rate.

The size of the compressor is based on the volumetric flow rate of working fluid. The vapour volume per unit latent heat of condensation (VPH) and the compression ratio (CR) together determine the power of the motor required to drive the compressor.

The volumetric flow rate of the working fluid into the compressor can be determined from equation

$$F = Q_{EV}(VPH) \qquad (2.10)$$

VPH values for seven working fluids at various condensing temperatures T_{CO} are listed in Table 2.1. Holland *et al.* [1982] have published extensive thermodynamic data for 21 working fluids including the seven listed in Table 2.1.

Table 2.1 Values of vapour volume per unit latent heat of condensation (VPH) for seven working fluids at various condensation temperatures T_{CO}

$T_{CO}(°C)$	VPH $(m^3\ MJ^{-1})$						
	R718	R600	R114	R600a	R717	R12	R290
140	0.191 68	0.066 32					
130	0.307 47	0.068 65					
120	0.405 11	0.075 39	0.082 72	0.075 627	0.0213 71		
110	0.542 63	0.085 70	0.091 81	0.0796 11	0.0217 99		
100	0.741 28	0.098 41	0.105 84	0.0877 73	0.0244 73		
90	1.034 74	0.114 68	0.123 70	0.0994 28	0.0283 02		
80	1.476 73	0.135 03	0.146 39	0.1149 33	0.0334 00		0.0618 135
70	2.162 63	0.162 01	0.176 46	0.1350 49	0.0400 07	0.0836 33	0.0681 491
60	3.257 22	0.196 92	0.209 21	0.1606 09	0.0485 91	0.0978 76	0.0778 531
50	5.052 29	0.243 18	0.264 20	0.1949 24	0.0601 77	0.1166 04	0.0909 471
40	8.125 45	0.366 12	0.333 00	0.2398 94	0.0753 94	0.1412 83	0.1081 830

Table 2.2 Values of condensation pressure P_{CO} for various condensation temperatures up to $T_{CO} = 100°C$

$T_{CO}(°C)$	P_{CO} (bar)				
	R600	R600a	R12	R290	R717
0	1.032 7	1.565 5	3.086 1	4.738 0	4.302 51
5	1.240 7	1.860 3	3.625 5	5.503 0	5.166 57
10	1.485 7	2.196 4	4.233 0	6.355 9	6.160 11
15	1.761 0	2.577 2	4.913 7	7.302 7	7.295 38
20	2.075 8	3.005 1	5.672 9	8.349 6	8.584 02
25	2.428 6	3.481 9	6.516 1	9.503 0	10.040 06
30	2.839 3	4.014 8	7.449 0	10.769 3	11.679 53
35	3.271 0	4.609 5	8.477 2	12.155 2	13.517 27
40	3.779 1	5.265 2	9.606 5	13.667 8	15.564 46
45	4.348 7	5.980 4	10.843 1	15.314 4	17.835 29
50	4.967 1	6.785 4	12.193 2	17.102 7	20.348 45
55	5.656 5	7.666 7	13.663 0	19.040 6	23.122 24
60	6.407 6	8.629 7	15.259 0	21.137 0	26.171 93
65	7.226 2	9.608 8	16.988 0	23.401 0	29.511 88
70	8.126 3	10.803 3	18.858 0	25.842 9	33.159 54
75	9.098 8	12.037 8	20.874 0	28.474 1	37.137 24
80	10.167 9	13.372 1		31.307 6	41.474 74
85	11.309 2	14.814 4		34.359 0	46.189 46
90	12.543 6	16.369 2		37.810 0	51.286 41
95	13.883 3	18.043 7		41.250 0	56.770 60
100	15.310 2	19.843 9			62.688 40

The only working fluids listed in Table 2.1 which have environmentally adverse ozone and global warming potentials are R12 and R114. The VPH value should be as low as possible in order to keep down the size and cost of the equipment. Although water (R718) is an excellent working fluid for high temperature heat pumps, its VPH values are too high for it to be economic at the lower temperatures.

Another important factor to consider in the selection of a working fluid is the condensation pressure P_{CO}. The P_{CO} values for five of the seven working fluids listed in Table 2.1 are listed in Table 2.2 up to condensation temperature values $T_{CO} = 100°C$. The P_{CO} values for ammonia (R717) are far too high for it to be considered for heat pumps operating at higher temperatures in spite of its favourable VPH values. High pressures require heavy-duty equipment and high VPH values require large units. Both these characteristics increase the capital cost of a heat pump system.

Table 2.3 lists the seven working fluids listed in Table 2.1 together with their safety classifications. Each working fluid listed in Table 2.3 is designated by two safety code numbers. The first is the American National Refrigeration Safety Code which defines three group of materials:

- Group 1 fluids are the safest working fluids. They have low toxicity and low flammability.

Table 2.3 Safety classifications for seven working fluids

Code number	Chemical formula	Critical temperature (°C)	Safety group	Safety class
R718	H_2O	373.0	–	–
R600	$CH_3CH_2CH_2CH_3$	152.0	3	5
R114	$CCl F_2CCl F_2$	145.7	1	6
R600a	$CH(CH_3)_3$	135.0	3	5b
R717	NH_3	132.2	2	2
R12	CCl_2F_2	112.0	1	6
R290	$CH_3CH_2CH_3$	97.0	3	5b

- Group 2 fluids are toxic and mildly flammable fluids.
- Group 3 fluids are flammable working fluids usually with an explosion risk.

The second is the American National Board of Fire Underwriters Refrigerant Toxicity Classification which defines six classes decreasing in toxicity from Class 1 (highly toxic) to Class 6 (non-toxic). Where the fluid has not been officially classified the entry is left blank.

2.6   The Choice of Compressor

The compressor is the heart of a mechanical vapour compression heat pump system [Sauer and Howell, 1983]. The selection of the most suitable compressor depends on the choice of working fluid and the size and capacity of the system.

The two main types are reciprocating piston compressors and rotary compressors. The latter includes sliding vane, lobe and centrifugal compressors. Figure 2.6 shows schematically various types of compressors used in heat pump systems.

Reciprocating compressors, driven by an electric motor, are the ones most commonly used. They are available in the following three types.

- The first is the open type in which the input drive shaft enters the compressor casing which contains the working fluid via a seal. The drive can be by an electric motor, an internal combustion engine or any other form of mechanical energy.
- The second is the semi-hermetic type in which the compressor and drive motor are sealed within a common housing. However, the casing is bolted to facilitate servicing and repair.
- The third type is the hermetic type in which the compressor and drive motor are sealed for life within a common housing, which is welded.

The semi-hermetic and hermetic compressors have fewer leakage, alignment and vibration problems than open type compressors. An additional

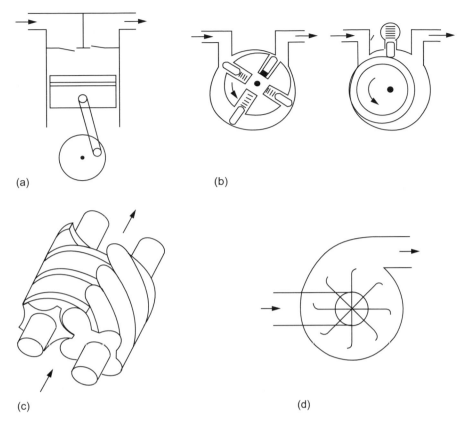

Figure 2.6 Schematic diagram of various types of compressors used in heat pump
systems: (a) reciprocating; (b) rotary vane; (c) screw; (d) centrifugal.

advantage is that the motor is located in the suction gas stream thus providing
cooling for the motor. However they have the disadvantage that they can only
be driven by an electric motor. Reciprocating compressors are unable to
handle wet gas, since liquid carry over can cause severe wear on valves,
pistons and rings.

The most important operating parameters are the volumetric efficiency η_{vol}
defined as

$$\eta_{vol} = \frac{\text{Actual volume flow of vapour}}{\text{Theoretical volume flow}} \tag{2.11}$$

and the overall efficiency η_{ov} defined as

$$\eta_{ov} = \frac{\text{Isentropic work}}{\text{Work input}} \tag{2.12}$$

For an open compressor, the work input is the shaft work. For a semi-hermetic compressor, the work in put is the electric power consumed by the motor.

The choice of a suitable lubricant is of significant importance for the correct operation of a reciprocating piston compressor. It is essential to maintain the minimum viscosity for safe lubrication under all the operating conditions. Additionally, the oil/working fluid system should be maintained chemically and thermally stable.

On start-up, the oil is transformed into a fine mist which is mixed with the working fluid. This mixture is carried through the entire system. It is therefore important that the pipelines are sized and arranged to facilitate the flow of oil returning to the compressor crankcase.

In order to control the capacity of the compressor, hot gas bypass or cylinder unloading is commonly used. Cylinder unloading employs a mechanism which automatically holds open some cylinder suction valves when less than full capacity is required.

Reciprocating piston compressors range in size from a few litres per hour to over $1000 \, \text{m}^3 \, \text{h}^{-1}$ with compression ratios up to 10.

Rotary vane, also called sliding vane, compressors are small capacity units with power inputs of less than 5 kW. They usually operate at relatively low pressures and low compression ratios. Rotary vane compressors have the advantage of operating with high volumetric efficiencies. A film of oil is maintained inside the cylinder and the vanes slide over this surface. Any breakdown of the oil film can result in damage to the sealing surfaces. The greater the number of blades, the higher the volumetric efficiency and the greater the frictional losses. Since the inlet flow is continuous, no valves are required. Rotary compressors have a better ability to withstand liquid slugging than reciprocating compressors. However, there is a greater possibility of gas leakage from the high-pressure to the low-pressure side in rotary compressors than is the case with reciprocating compressors.

The most common type of screw compressor [Reay and Macmichael, 1979] consists of two rotors meshing together inside a sleeve. The male rotor has a number of semicircular lobes formed in a helix along the body of the rotor. The female rotor has a corresponding number of cavities which form the opposite helix. Compression is achieved in a volume within one channel of the female rotor. Discharge occurs through a part in the end of the casing. All the vapour is discharged. Some screw compressors are designed to run oil free. However, most screw compressors are of the oil injection type in which the oil is continuously pumped into the suction volume.

The oil provides a tight seal between the rotors and the casing. However, the oil needs to be separated after discharge. In oil free screw compressors, there are sealing problems and the pressures are limited.

Screw compressors can operate at relatively high pressures and at compression ratios up to 12. However, they are relatively expensive and noisy compared with reciprocating compressors.

Centrifugal compressors are normally high-capacity units with limited compression ratios less than 4.5. In a centrifugal compressor, a rotating impeller delivers kinetic energy to the vapour which is converted into pressure energy. The greater the rotational speed of the impeller, the higher the pressure developed. Other impellers can be mounted in series on the same shaft in multistage compressor units to produce higher pressures. Impeller speeds normally range from 3000 to 20 000 rpm. Centrifugal compressors can operate with high efficiencies of about 80–85%.

In any system using steam, oil free compressors are required. EA Technology Ltd, Capenhurst, Chester, UK, have developed a relatively low-cost unit for steam compression. For this purpose, they modified a Roots rotary twin-lobe type blower and incorporated a novel water-cooled seal design. Compression ratios are limited to two. However, it is possible to employ multistage units to obtain a lift of about 1 bar per stage. The units can be operated up to a temperature of 150°C with water injection for desuperheating. At subatmospheric pressures, the steam throughput is limited to 2500 kg h^{-1} because of the large volumes of vapour involved.

2.7   Compressor Drive Units

Compressors can be driven by any kind of power producing unit such as: (a) electric motors; (b) internal and external combustion engines; (c) gas driven turbines; (d) water driven turbines; (e) compressed air driven motors, and even (f) wind driven units. However, electric motors are by far the most commonly used power units and have the important advantage of many years of use. Consequently, they are normally relatively inexpensive, readily available, very reliable, and require little maintenance. They have a high efficiency based on the electric energy supplied.

However, if the electric energy is generated relatively inefficiently, electric motor driven heat pumps can be significantly less efficient than engine driven units in terms of the primary energy ratio defined as

$$(PER) = \frac{\text{Useful heat delivered}}{\text{Primary energy input}} \qquad (2.13)$$

Potentially, engine driven heat pumps can be very attractive, from a primary energy point of view, since the heat from the engine coolant can be used to supplement the heat produced in the condenser at the heat pump. In a gas engine, about 30% of the input energy is rejected into the cooling water. Most of this can be recovered and added to the heat produced by the heat pump.

However, engines tend to be relatively expensive compared with electric motors. They also require more maintenance and are generally less reliable.

It is possible that, eventually, engine-driven heat pumps, and in particular gas engine systems, will have benefited sufficiently by the learning experience to compete with electric motors in terms of cost, reliability, low maintenance, etc.

However, this is unlikely to occur until energy conservation becomes a top priority and the potential for engine-driven heat pumps to operate with high primary energy ratios becomes widely appreciated.

2.8   References

Devotta, S. (1995) Alternative heat pump working fluids to CFCs, *Heat Recovery Systems and CHP*, **15**, (3), 273–9.

Holland, F. A., Watson, F. A. and Devotta, S. (1982) *Thermodynamic Design Data for Heat Pump Systems*, Pergamon Press, Oxford, UK.

Reay, D. A. and Macmichael, D. B. A. (1979) *Heat Pumps Design and Applications*, Pergamon Press, Oxford, UK.

Sauer, H. J. and Howell, R. H. (1983) *Heat Pump Systems*, John Wiley, New York, USA.

3 Design basis for absorption heat pumps operating with water/salt working pairs

3.1 Nomenclature

(COP)	coefficient of performance [dimensionless]
CR	compression ratio [dimensionless]
FR	flow ratio [dimensionless]
H	enthalpy per unit mass [kJ kg^{-1}]
HPE	heat pump effectiveness [dimensionless]
M	mass flow rate [kg s^{-1}]
P	absolute pressure [kPa]
$\bar{P}$	vapour pressure [kPa]
Q	heat flow rate [kW] or heat quantity [kWh]
T	temperature [°C or K]
W	rate of work delivered to the shaft of the compressor [kW] or quantity of work [kWh]
x	mole fraction of salt [dimensionless]
X	concentration of salt by weight [%]

Subscripts

A	actual
$\bar{A}$	absolute ambient
AB	absorber
C	Carnot
CO	condenser
EH	enthalpic
EV	evaporator
GE	generator
H	heating
L	lower
R	refrigerant
S	solution
SL	salt
W	water

3.2   Introduction

Absorption heat pumps are devices for raising the temperature of low-grade heat to a more useful level using a smaller amount of relatively high-grade heat energy. The higher the absolute temperature T of a quantity of heat Q, the lower its entropy Q/T and the greater its quality as measured by its exergy or work equivalent W given by the equation

$$W = Q\left(1 - \frac{T_{\bar{A}}}{T}\right) \tag{3.1}$$

with reference to an absolute ambient temperature $T_{\bar{A}}$.

A basic heat pump cycle can be thought of as a heat engine operating in reverse. Figure 3.1 shows a schematic comparison of a simple heat engine and a simple heat pump operating between two temperature levels T_H and T_L. A heat-driven absorption heat pump consists of a primary and secondary circuit. The primary circuit is a basic heat pump cycle and the secondary circuit can be thought of as a basic heat engine cycle which converts the heat supplied to drive the overall system into work to operate the heat pump cycle. As shown by equation (3.1), the higher the temperature of the heat supplied, the greater is its capacity to provide the required work input to the basic heat pump cycle.

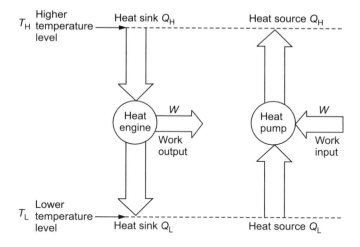

Figure 3.1 Schematic diagram of a heat engine and a heat pump operating between the same temperature levels.

3.3   The Carnot Coefficient of Performance of an Absorption Heat Pump

The primary circuit of a heat driven absorption heat pump is identical to that of a work driven mechanical vapour compression heat pump as illustrated in

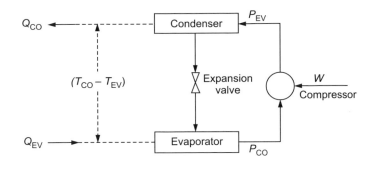

(a)

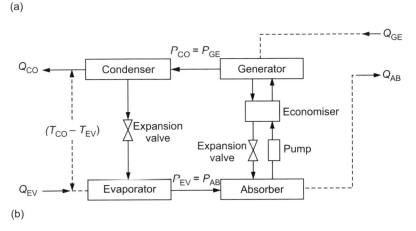

(b)

Figure 3.2 Schematic comparison of the primary circuits in (a) compressor driven vapour compression and (b) absorption heat pump cycles.

Figure 3.2. In both cycles, the pure working fluid or refrigerant is condensed in the condenser producing a quantity of heat Q_{CO} at an absolute temperature T_{CO} with an entropy value of Q_{CO}/T_{CO}. After passing through the expansion valve and having its pressure reduced from P_{CO} to P_{EV}, the refrigerant is evaporated in the evaporator absorbing a quantity of heat Q_{EV} at an absolute temperature T_{EV} with an entropy value of Q_{EV}/T_{EV}. For thermodynamically reversible processes in the condenser and evaporator, the reduction in entropy in the condenser would equal the gain in entropy in the evaporator so that

$$\frac{Q_{CO}}{T_{CO}} = \frac{Q_{EV}}{T_{EV}} \tag{3.2}$$

An energy balance on the compressor driven system shown in Figure 3.2. gives

$$Q_{CO} = Q_{EV} + W \tag{3.3}$$

where W is the input of mechanical energy to the compressor required to produce the compression ratio

$$(CR) = \frac{P_{CO}}{P_{EV}} \tag{3.4}$$

The coefficient of performance (COP) of a mechanical vapour compression heat pump is the ratio

$$(COP) = \frac{Q_{CO}}{W} \tag{3.5}$$

Equation (3.5) can also be written as

$$(COP) = \frac{Q_{CO}}{(Q_{CO} - Q_{EV})} \tag{3.6}$$

For a thermodynamically reversible system in which there is no increase in entropy, an ideal Carnot coefficient of performance $(COP)_C$ can be obtained by substituting equation (3.2) into equation (3.6) to give

$$(COP)_C = \frac{T_{CO}}{(T_{CO} - T_{EV})} \tag{3.7}$$

which is the ratio of the condensing temperature T_{CO} of the refrigerant to the gross temperature lift $(T_{CO} - T_{EV})$ produced by the heat pump. The Carnot coefficient of performance $(COP)_C$ depends only on the temperatures in the system and is independent of the working fluid or refrigerant used. It can never be achieved in practice. In the heat driven type of absorption heat pump shown in Figure 3.2, the role of the compressor is performed by a secondary circuit which is used to produce the pressure ratio $(CR) = P_{CO}/P_{EV}$. In the secondary circuit of an absorption heat pump operating with a water/salt working pair, water is the refrigerant which is recycled in the primary circuit whilst an aqueous salt solution is recycled in the secondary circuit, which consists of an absorber, a generator (also known as a desorber), an economiser, a pump and an expansion valve.

A quantity of heat Q_{GE} at an absolute temperature T_{GE}, with an entropy value of Q_{GE}/T_{GE}, is supplied to the generator to separate or desorb pure water vapour from the concentrated salt solution in the generator. The pure water vapour desorbed in the generator at a pressure $P_{GE} \approx P_{CO}$ then flows from the generator to the condenser as the concentrated salt solution flows from the generator through the economiser and the expansion valve to the absorber at a pressure $P_{AB} \cong P_{EV}$. The pure water vapour from the evaporator is then absorbed in the concentrated salt solution in the absorber, releasing a quantity of heat Q_{AB} at an absolute temperature T_{AB} with an entropy value of Q_{AB}/T_{AB}. The diluted salt solution is then pumped through the economiser back to the absorber to complete the cycle. A heat balance on the cycle gives

$$Q_{GE} + Q_{EV} = Q_{AB} + Q_{CO} \tag{3.8}$$

In an absorption heat pump, the diluted salt solution which is pumped from the absorber to the generator is conventionally referred to as the refrigerant strong solution. On the other hand, the concentrated salt solution which flows from the generator to the absorber, through the expansion valve, is conventionally referred to as the refrigerant poor or weak solution. For a thermodynamically reversible absorption heat pump cycle, the net change in entropy is zero, so that

$$\frac{Q_{GE}}{T_{GE}} + \frac{Q_{EV}}{T_{EV}} = \frac{Q_{AB}}{T_{AB}} + \frac{Q_{CO}}{T_{CO}} \tag{3.9}$$

Substitution of equation (3.2) into equation (3.9) gives

$$\frac{Q_{GE}}{T_{GE}} = \frac{Q_{AB}}{T_{AB}} \tag{3.10}$$

The coefficient of performance (COP) of an absorption heat pump is the ratio of the sum of the heat produced in the absorber and condenser $Q_{AB} + Q_{CO}$ to the heat supplied to the generator Q_{GE}

$$(COP)_H = \frac{Q_{AB} + Q_{CO}}{Q_{GE}} \tag{3.11}$$

Normally, the temperature of the refrigerant in the condenser is equal to the temperature in the absorber, i.e. $T_{CO} = T_{AB}$. Substitute equation (3.8) into equation (3.11) to give

$$(COP)_H = \frac{Q_{GE} + Q_{EV}}{Q_{GE}} = 1 + \frac{Q_{EV}}{Q_{GE}} \tag{3.12}$$

When $T_{AB} = T_{CO}$, equation (3.9) becomes

$$\frac{Q_{GE}}{T_{GE}} + \frac{Q_{EV}}{T_{EV}} = \frac{Q_{AB} + Q_{CO}}{T_{CO}} \tag{3.13}$$

Substitute equation (3.8) into equation (3.13) to give

$$\frac{Q_{GE}}{T_{GE}} + \frac{Q_{EV}}{T_{EV}} = \frac{Q_{GE} + Q_{EV}}{T_{CO}} \tag{3.14}$$

which can be rewritten in the form

$$\frac{Q_{GE}}{T_{GE}} = \left(1 - \frac{T_{CO}}{T_{GE}}\right)\left(\frac{T_{EV}}{T_{CO} - T_{EV}}\right) \tag{3.15}$$

Substitute equation (3.15) in equation (3.12) to give

$$(COP)_{CH} = \left(1 - \frac{T_{EV}}{T_{GE}}\right)\left(\frac{T_{CO}}{T_{CO} - T_{EV}}\right) \tag{3.16}$$

Equation (3.16) is the ideal Carnot coefficient of performance for a thermo-dynamically reversible absorption heat pump cycle. A comparison of equations (3.7) and (3.16) shows that the Carnot coefficient of performance of a thermodynamically reversible absorption heat pump is less than the Carnot coefficient of performance of a thermodynamically mechanical vapour compression heat pump by a factor which is the Carnot cycle efficiency of a thermodynamically reversible heat engine taking in heat at an absolute temperature T_{GE} and rejecting it at an absolute temperature T_{EV}. Thus, thermodynamically, an absorption heat pump is equivalent to a heat engine driving a compressor driven vapour compression heat pump. Equation (3.16) shows that the theoretically ideal Carnot coefficient of performance $(COP)_{CH}$ is independent of the working pair and is dependent only on the temperatures $T_{EV}, T_{CO} = T_{AB}$ and T_{GE}.

In practice, the Carnot coefficient of performance can never be achieved because of the unavoidable irreversibilities that arise from the operation of a practical system. For example, the net temperature lift for a heat pump is less than the gross temperature lift $(T_{CO} - T_{EV})$ because of the required temperature difference driving forces in the heat exchangers.

3.4   The Enthalpic Coefficient of Performance of an Absorption Heat Pump

It is much more realistic to consider a heat pump cycle on the basis of the mass flow rates and enthalpies of the various streams. For the different state points of the absorption heat pump cycle shown in Figure 3.3, the following mass balances can be written

$$M_1 = M_6 = M_7 = M_8 = M_R \tag{3.17}$$

$$M_3 = M_5 = M_{AB} \tag{3.18}$$

$$M_5 = M_4 + M_6 \tag{3.19}$$

$$M_3 = M_1 + M_2 \tag{3.20}$$

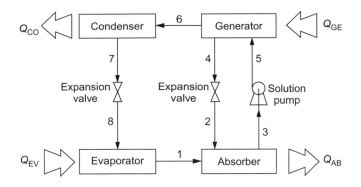

Figure 3.3 Schematic diagram of an absorption heat pump cycle showing various state points.

where M_R is the mass flow rate of the pure refrigerant and M_{AB} is the flow rate of the refrigerant rich solution being pumped from the absorber to the generator. An important design parameter in absorption heat pumps is the flow ratio (FR) defined as

$$(FR) = \frac{M_{AB}}{M_R} \qquad (3.21)$$

(FR) can also be written in terms of the % salt concentrations by weight in the generator X_{GE} and the absorber X_{AB} as

$$(FR) = \frac{X_{GE}}{X_{GE} - X_{AB}} \qquad (3.22)$$

In terms of the state points in Figure 3.3, the following heat balances can be written. For the condenser

$$Q_{CO} = M_6 H_6 - M_8 H_8 \qquad (3.23)$$

which can be rewritten in terms of equation (3.17) as

$$Q_{CO} = M_R (H_6 - H_8) \qquad (3.24)$$

For the absorber

$$Q_{AB} = M_1 H_1 + M_4 H_4 - M_5 H_5 \qquad (3.25)$$

which can be rewritten in terms of equations (3.17)–(3.21) as

$$Q_{AB} = M_R \{ H_1 + [(FR) - 1]H_4 - (FR)H_5 \} \qquad (3.26)$$

For the generator

$$Q_{GE} = M_6 H_6 + M_4 H_4 - M_5 H_5 \qquad (3.27)$$

which can be rewritten in terms of equations (3.17)–(3.21) as

$$Q_{GE} = M_R\{H_6 + [(FR) - 1]H_4 - (FR)H_5\} \qquad (3.28)$$

The coefficient of performance of an absorption heat pump used for heating can be written as

$$(COP)_H = \frac{Q_{AB} + Q_{CO}}{Q_{GE}} \qquad (3.11)$$

The coefficient of performance of an absorption heat pump used for heating can be written in terms of enthalpies by substituting equations (3.24), (3.26) and (3.28) into equation (3.11) to give the enthalpic coefficient of performance

$$(COP)_{EH} = \frac{H_1 + [(FR) - 1]H_4 - (FR)H_5 + H_6 - H_8}{H_6 + [(FR) - 1]H_4 - (FR)H_5} \qquad (3.29)$$

Equation (3.29) is commonly used as the basis for the evaluation of the performance of absorption heat pumps since it is the theoretical maximum obtainable coefficient of performance for an absorption heat pump heating system.

3.5   The Absorption Heat Pump Effectiveness Factor

The theoretical Carnot efficiencies for heat pumps and heat engines are unrealistic measures of performance. For example, a theoretical reversible heat engine would have a zero power output. A heat pump effectiveness HPE for an absorption heat pump used for heating can be defined as the ratio of the actual coefficient of performance $(COP)_A$ to the enthalpic coefficient of performance $(COP)_{EH}$ by the equation

$$HPE = \frac{(COP)_A}{(COP)_{EH}} \qquad (3.30)$$

The use of the HPE factor is a realistic and practical measure for the evaluation of heat pump systems [Chaudhari et al., 1985].

3.6   Raoult's Law Deviations for Aqueous Salt Solutions

For an aqueous salt solution which obeys Raoult's law, the vapour pressure P of the solution at a particular temperature is given by the equation

$$P = x\overline{P_{SL}} + (1 - x)\overline{P_w} \tag{3.31}$$

where x is the mole fraction of the salt in the solution, $\overline{P_w}$ is the vapour pressure of pure water and $\overline{P_{SL}}$ is the vapour pressure of the pure salt. Since $\overline{P_{SL}}$ is negligible, equation (3.31) can be written as

$$P = (1 - x)\overline{P_w} \tag{3.32}$$

which can be written in the form

$$x = 1 - \frac{P}{\overline{P_W}} \tag{3.33}$$

For an aqueous salt solution which obeys Raoult's law, the mole fraction of salt x_{GE} in the generator of an absorption heat pump is given by equation (3.33) written as

$$x_{GE} = 1 - \frac{\left(\overline{P_W}\right)_{at\, T_{CO}}}{\left(\overline{P_W}\right)_{at\, T_{GE}}} \tag{3.34}$$

where $\left(\overline{P_W}\right)_{at\, T_{CO}}$ is the vapour pressure of water at the condenser temperature T_{CO} and $\left(\overline{P_W}\right)_{at\, T_{GE}}$ is the vapour pressure of water at the generator temperature T_{GE}. For the negligible pressure drop between the generator and the condenser $\left(\overline{P_W}\right)_{at\, T_{CO}} \cong \left(\overline{P_W}\right)_{at\, T_{GE}}$ so that equation (3.34) can be written as

$$x_{GE} = 1 - \frac{P_{CO}}{\left(\overline{P_W}\right)_{at\, T_{GE}}} \tag{3.35}$$

and equation (3.32) can be written as

$$P_{CO} = (1 - x_{GE})\left(\overline{P_W}\right)_{at\, T_{GE}} \tag{3.36}$$

for a solution which obeys Raoult's law. Aqueous solutions of lithium bromide exhibit large negative deviations from Raoult's law as illustrated in Figure 3.4. This makes these solutions particularly attractive for use in absorption heat pump systems.

Solutions which have a negative deviation from the Raoult's law have a lower vapour pressure for a given mole fraction of water $(1 - x)$ than that predicted by Raoult's law. Thus for a given solution vapour pressure, the mole fraction of salt x is less and the mole fraction of water $(1 - x)$ is greater than for an ideal solution. Since the mole fraction x_{GE} in the generator is less in the case of lithium bromide than for an ideal solution, the water concentration $(1 - x_{GE})$ is greater for a given value of P_{CO} in equation (3.36). Thus a higher

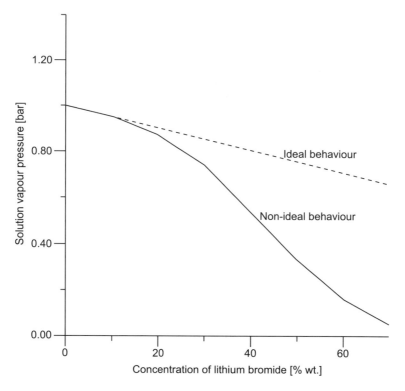

Figure 3.4 Plot of vapour pressure against concentration for lithium bromide solutions at 100°C showing negative deviation from Raoult's law.

condenser temperature T_{CO} can be achieved for a given generator temperature T_{GE} since a lower salt concentration x_{GE} is required.

Solutions which have a negative deviation from Raoult's law have the great advantage of reducing the volume flow of the solution in the secondary circuit for a given flow rate of the water refrigerant in the primary circuit. They have a lower vapour pressure for a given mole fraction of water $(1 - x)$ than that predicted by Raoult's law. This in turn reduces the capital cost of the equipment.

3.7   Pressure/Temperature Plots for Absorption Heat Pump Systems

In a conventional absorption heat pump there are two pressure levels

$$P_{CO} = P_{GE} > P_{EV} = P_{AB}$$

and either three or four temperature levels

$$T_{\text{GE}} = T_{\text{CO}} > T_{\text{AB}} = T_{\text{EV}}$$

depending on whether the condenser and absorber are operated at the same temperature or not. The normal convention is to show the absorption cycle on a pressure against temperature plot as in Figure 3.5. Although, since the temperatures are the most important factors for the design engineer to consider, a plot of temperature against pressure as in Figure 3.6 may be more appropriate.

Figure 3.7 is a plot of the logarithm of the vapour pressure against the inverse of the temperature for a water/salt solution. It can be seen that at the evaporator pressure P_{EV}, the pure water at the evaporator temperature T_{EV} is in equilibrium with the salt solution at the higher absorber temperature T_{AB}. When the pressure in the absorber is slightly less than in the evaporator, the water vapour will evaporate at the temperature T_{EV} and be absorbed at the significantly higher absorber temperature T_{AB} so that heat is being pumped through a temperature lift $(T_{\text{AB}} - T_{\text{EV}})$. This is the essential principle of the absorption heat pump.

Of course, if the existing salt solution remained in the absorber, its concentration would be diluted by the condensing water and the absorber temperature T_{AB} and the temperature lift $(T_{\text{AB}} - T_{\text{EV}})$ would progressively decrease. In a continuously operating absorption heat pump, the concentration and the temperature T_{AB} in the absorber are kept constant by continuously pumping the solution to a higher pressure in the generator where the excess water is driven off by an input of relatively high grade heat Q_{GE}. The concentrated salt solution, which is now weak in refrigerant, is reduced in pressure through an expansion valve and returned to the absorber.

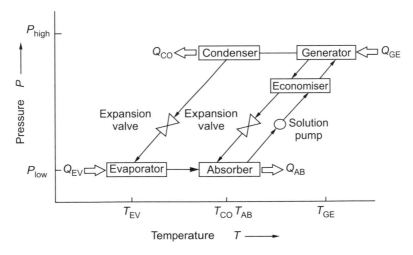

Figure 3.5 Schematic diagram of an absorption heat pump cycle in a plot of pressure P against temperature T.

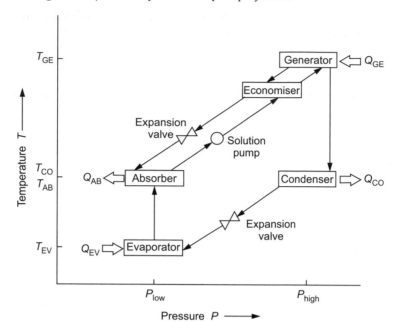

Figure 3.6 Schematic diagram of an absorption heat pump cycle in a plot of temperature T against pressure P.

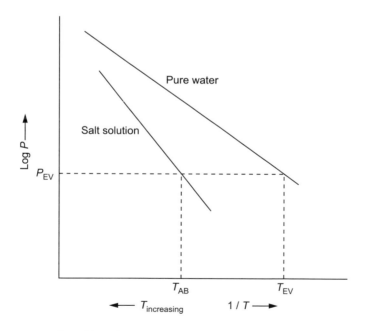

Figure 3.7 Plot of log of vapour pressure against inverse of temperature for a water/salt solution.

Water/salt systems have the advantage that water has a high latent heat of vaporisation (2256 kJ kg^{-1} at the normal boiling point of 100°C). However water also has the disadvantage of a relatively low vapour pressure. Therefore it has a high specific volume which requires large pieces of equipment. A further disadvantage is that heat pumps which use water/salt systems cannot operate below 0°C, the freezing temperature of water, unless an antifreeze agent is added.

3.8 Calculated Enthalpic Coefficients of Performance for the Water/Lithium Bromide and the Water/Carrol Systems

In the context of Figure 3.3 the following assumptions were considered as the theoretical basis for the experimental programme.

1. There was thermodynamic equilibrium throughout the entire system.
2. The analysis was made under steady-state conditions.
3. A rectifier was not used because the absorbent does not evaporate in the range under consideration.
4. The solution leaving the generator and the absorber was saturated. In the condenser and the evaporator the working fluid was saturated.
5. Heat losses and pressure drops in the tubing and the components were considered negligible.
6. The flow through the valves was isenthalpic.

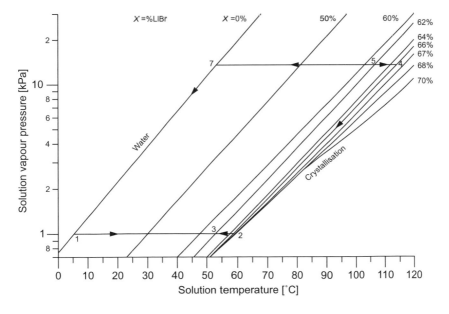

Figure 3.8 Absorption cycle on an equilibrium chart for the water/lithium bromide system.

The temperatures at the exits of the main items of equipment T_1, T_2, T_5, T_6 and T_7 and the heat load Q_{GE} in the generator are all known.

The different state points of the cycle, the theoretical Carnot coefficient of performance $(COP)_{CH}$, the enthalpy based coefficient of performance $(COP)_{EH}$, the concentrations of the solution in the absorber and the generator and the flow ratio for the water/lithium bromide and water/Carrol were determined using the following procedure in the context of the thermodynamic data available in the literature [Gupta and Sharma, 1976; McNeely, 1979; Reimann and Biermann, 1984; Eisa et al., 1986; Eisa and Holland, 1987; Khoeler et al., 1987; Macriss and Zawaki, 1989; Moran and Herold, 1989; Patil et al., 1989; Di Guillio et al., 1990; Lee et al., 1990; Best et al., 1995] (Figures 3.8–3.12).

From assumption 1:

$$T_6 = T_5 \tag{3.37}$$

Since the working fluid is considered saturated leaving the condenser and the evaporator, then the pressure at those state point may be obtained by the following equations.

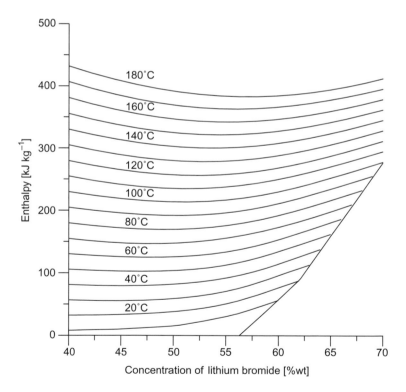

Figure 3.9 Plot of enthalpy per unit mass against concentration for water/lithium bromide solutions at various temperatures.

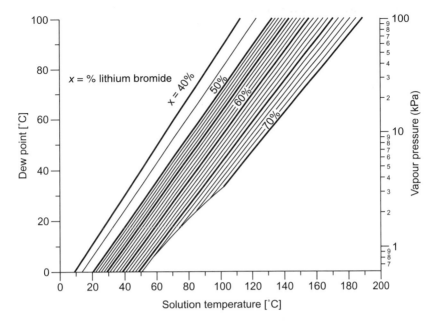

Figure 3.10 Plot of dew point and vapour pressure against temperature for water/
lithium bromide solutions at various concentrations.

$$P_1 = P(T_1) \tag{3.38}$$

and

$$P_7 = P(T_7) \tag{3.39}$$

Since the pressure drops in the tubing and the components are considered
negligible then

$$P_1 = P_2 = P_3 = P_8 \tag{3.40}$$

and

$$P_4 = P_5 = P_6 = P_7 \tag{3.41}$$

From assumption 4, and using equation (3.41), the concentration of the
refrigerant-poor solution leaving the generator can be estimated as

$$X_5 = X(P_5, T_5) \tag{3.42}$$

In the same way, the concentration of the refrigerant-rich solution leaving the
absorber can be estimated as

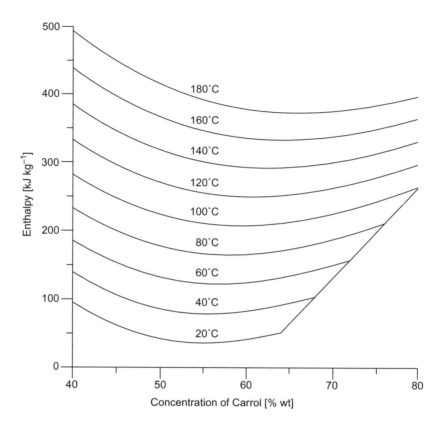

Figure 3.11 Plot of enthalpy per unit mass against concentration for water/Carrol solutions at various temperatures.

$$X_2 = X(P_2, T_2) \tag{3.43}$$

Since there is no mass transfer between the generator and the absorber then

$$X_5 = X_3 \tag{3.44}$$

and

$$X_2 = X_4 \tag{3.45}$$

From assumption 3, the concentration of the absorbent in the vapour leaving the generator is zero, then

$$X_1 = X_6 = X_7 = X_8 = 0 \tag{3.46}$$

The enthalpies at the exit of the generator and the absorber can be estimated as

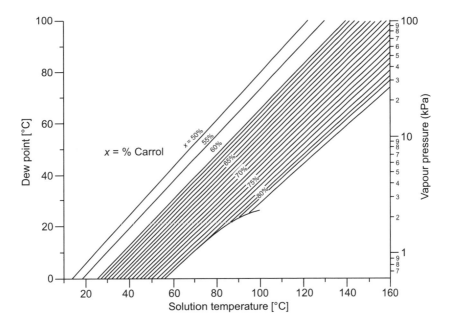

Figure 3.12 Plot of dew point and vapour pressure against temperature for water/Carrol solutions at various concentrations.

$$H_5 = H(T_5, X_5) \tag{3.47}$$

and

$$H_2 = H(T_2, X_2) \tag{3.48}$$

On the basis of assumption 4, the enthalpies at the exit of the condenser and the evaporator can be estimated as

$$H_7 = H(T_7) \tag{3.49}$$

and

$$H_1 = H(T_1) \tag{3.50}$$

Considering that the vapour upon leaving the generator is superheated

$$H_6 = H(P_6, T_6) \tag{3.51}$$

Since the absorbent does not evaporate in the temperature range under consideration, from equations (3.21) and (3.22), the flow ratio (FR) can be written for this system as

$$FR = \frac{M_{AB}}{M_R} = \frac{X_5}{X_5 - X_2} \qquad (3.52)$$

To analyse the behaviour of an absorption heat pump operating with the water/lithium bromide and water/Carrol mixtures, the main temperatures of the system were assumed and the amount of heat delivered to the generator was taken as 1 kJ. The temperature ranges considered for the analysis were

Table 3.1 Derived thermodynamic design data for absorption heat pump systems operating on water/lithium bromide for heating

T_{EV} (°C)	T_{AB} (°C)	T_{CO} (°C)	T_{GE} (°C)	X_{AB} (% by weight)	X_{GE} (% by weight)	FR (dimension-less)	$(COP)_{EH}$ (dimension-less)	$(COP)_{CH}$ (dimension-less)
60.0	75.0	85.0	110.0	42.55	47.13	10.29	1.70	2.22
60.0	75.0	85.0	115.0	42.55	49.76	6.90	1.74	2.37
60.0	75.0	85.0	120.0	42.55	52.28	5.37	1.76	2.53
60.0	75.0	85.0	125.0	42.55	54.65	4.52	1.77	2.67
60.0	75.0	90.0	110.0	42.55	44.18	27.10	1.48	2.01
60.0	75.0	90.0	115.0	42.55	46.86	10.87	1.66	2.14
60.0	75.0	90.0	120.0	42.55	49.46	7.16	1.71	2.27
60.0	75.0	90.0	125.0	42.55	51.97	5.52	1.74	2.39
60.0	75.0	95.0	115.0	42.55	43.97	30.96	1.42	1.98
60.0	75.0	95.0	120.0	42.55	46.59	11.53	1.62	2.09
60.0	75.0	95.0	125.0	42.55	49.17	7.43	1.68	2.20
65.0	75.0	85.0	110.0	38.50	43.26	9.09	1.72	2.54
65.0	75.0	85.0	115.0	38.50	45.33	6.64	1.75	2.74
65.0	75.0	85.0	120.0	38.50	47.22	5.42	1.77	2.94
65.0	75.0	85.0	125.0	38.50	48.98	4.67	1.78	3.12
65.0	75.0	90.0	110.0	38.50	40.84	17.45	1.58	2.24
65.0	75.0	90.0	115.0	38.50	43.20	9.19	1.69	2.39
65.0	75.0	90.0	120.0	38.50	45.15	6.79	1.73	2.55
65.0	75.0	90.0	125.0	38.50	47.03	5.51	1.74	2.70
65.0	75.0	95.0	115.0	38.50	40.74	18.19	1.53	2.16
65.0	75.0	95.0	120.0	38.50	42.96	9.63	1.65	2.29
65.0	75.0	95.0	125.0	38.50	44.98	6.94	1.70	2.42
65.0	85.0	85.0	120.0	45.40	47.22	25.95	1.50	2.51
65.0	85.0	85.0	125.0	45.40	48.98	13.68	1.62	2.70
65.0	85.0	90.0	125.0	45.40	47.03	28.85	1.44	2.36
70.0	75.0	85.0	110.0	38.50	43.26	9.09	1.72	3.09
70.0	75.0	85.0	115.0	38.50	45.33	6.64	1.76	3.36
70.0	75.0	85.0	120.0	38.50	47.22	5.42	1.77	3.62
70.0	75.0	85.0	125.0	38.50	48.98	4.67	1.78	3.87
70.0	75.0	90.0	110.0	38.50	40.84	17.45	1.58	2.57
70.0	75.0	90.0	115.0	38.50	43.20	9.19	1.69	2.77
70.0	75.0	90.0	120.0	38.50	45.15	6.79	1.73	2.96
70.0	75.0	90.0	125.0	38.50	47.03	5.51	1.75	3.15
70.0	75.0	95.0	115.0	38.50	40.74	18.19	1.54	2.41
70.0	75.0	95.0	120.0	38.50	42.96	9.63	1.65	2.57
70.0	75.0	95.0	125.0	38.50	44.98	6.94	1.70	2.72

$$60\,°C \le T_{EV} \le 76\,°C$$

$$76\,°C \le T_{AB} \le 115\,°C$$

$$85\,°C \le T_{CO} \le 95\,°C$$

$$110\,°C \le T_{GE} \le 125\,°C$$

where the generator and condenser temperatures were increased in 5°C increments, and the evaporator and absorber temperatures in 10°C increments and the $(COP)_{EH}$ values were calculated from equation (3.29)

Table 3.2 Derived thermodynamic design data for absorption heat pump systems operating on water/lithium bromide for heating

T_{EV} (°C)	T_{AB} (°C)	T_{CO} (°C)	T_{GE} (°C)	X_{AB} (% by weight)	X_{GE} (% by weight)	FR (dimension-less)	$(COP)_{EH}$ (dimension-less)	$(COP)_{CH}$ (dimension-less)
70.0	85.0	85.0	110.0	42.16	43.26	39.33	1.48	2.49
70.0	85.0	85.0	115.0	42.16	45.33	14.30	1.67	2.77
70.0	85.0	85.0	120.0	42.16	47.22	9.33	1.73	3.04
70.0	85.0	85.0	125.0	42.16	48.98	7.18	1.75	3.30
70.0	85.0	90.0	115.0	42.16	43.20	41.54	1.42	2.33
70.0	85.0	90.0	120.0	42.16	45.15	15.10	1.62	2.53
70.0	85.0	90.0	125.0	42.16	47.03	9.66	1.69	2.72
70.0	85.0	95.0	120.0	42.16	42.96	53.70	1.32	2.22
70.0	85.0	95.0	125.0	42.16	44.98	15.95	1.58	2.38
70.0	95.0	85.0	125.0	47.99	48.98	49.47	1.39	2.72
75.0	75.0	85.0	110.0	38.50	43.26	9.09	1.73	4.18
75.0	75.0	85.0	115.0	38.50	45.33	6.64	1.76	4.59
75.0	75.0	85.0	120.0	38.50	47.22	5.42	1.78	4.98
75.0	75.0	85.0	125.0	38.50	48.98	4.67	1.78	5.37
75.0	75.0	90.0	110.0	38.50	40.84	17.45	1.58	3.12
75.0	75.0	90.0	115.0	38.50	43.20	9.19	1.69	3.39
75.0	75.0	90.0	120.0	38.50	45.15	6.79	1.73	3.66
75.0	75.0	90.0	125.0	38.50	47.03	5.51	1.75	3.91
75.0	75.0	95.0	115.0	38.50	40.74	18.19	1.54	2.79
75.0	75.0	95.0	120.0	38.50	42.96	9.63	1.66	2.99
75.0	75.0	95.0	125.0	38.50	44.98	6.94	1.70	3.19
75.0	85.0	85.0	110.0	38.50	43.26	9.09	1.78	3.27
75.0	85.0	85.0	115.0	38.50	45.33	6.64	1.81	3.69
75.0	85.0	85.0	120.0	38.50	47.22	5.42	1.81	4.10
75.0	85.0	85.0	125.0	38.50	48.98	4.67	1.82	4.50
75.0	85.0	90.0	110.0	38.50	40.84	17.45	1.66	2.51
75.0	85.0	90.0	115.0	38.50	43.20	9.19	1.75	2.79
75.0	85.0	90.0	120.0	38.50	45.15	6.79	1.77	3.07
75.0	85.0	90.0	125.0	38.50	47.03	5.51	1.79	3.33
75.0	85.0	95.0	115.0	38.50	40.74	18.19	1.60	2.35
75.0	85.0	95.0	120.0	38.50	42.96	9.63	1.71	2.55
75.0	85.0	95.0	125.0	38.50	44.98	6.94	1.74	2.75
75.0	95.0	85.0	115.0	44.88	45.33	100.73	1.32	2.79

$$(COP)_{EH} = \frac{H_1 + [(FR) - 1]H_4 - (FR)H_5 + H_6 - H_8}{H_6 + [(FR) - 1]H_4 - (FR)H_5} \qquad (3.29)$$

The results of these calculations are listed in Tables 3.1 and 3.2 for the water/lithium bromide working pair and in Tables 3.3 and 3.4 for the water/Carrol working pair.

Values for the enthalpic coefficient of performance $(COP)_{EH}$ for the water/lithium bromide and water/Carrol working pairs and the correspond-

Table 3.3 Derived thermodynamic design data for absorption heat pump systems operating on water/Carrol for heating

T_{EV} (°C)	T_{AB} (°C)	T_{CO} (°C)	T_{GE} (°C)	X_{AB} (% by weight)	X_{GE} (% by weight)	FR (dimension-less)	$(COP)_{EH}$ (dimension-less)	$(COP)_{CH}$ (dimension-less)
60.0	75.0	90.0	110.0	45.8	49.0	15.4	1.64	2.01
60.0	75.0	90.0	115.0	45.8	52.7	7.6	1.72	2.14
60.0	75.0	90.0	120.0	45.8	55.8	5.6	1.74	2.27
60.0	75.0	90.0	125.0	45.8	58.5	4.6	1.75	2.39
60.0	75.0	95.0	115.0	45.8	48.6	17.2	1.56	1.98
60.0	75.0	95.0	120.0	45.8	52.4	7.9	1.69	2.09
60.0	75.0	95.0	120.0	45.8	55.5	5.7	1.72	2.20
60.0	75.0	100.0	120.0	45.8	48.3	19.5	1.50	1.95
60.0	75.0	100.0	125.0	45.8	52.1	8.3	1.65	2.05
60.0	80.0	90.0	115.0	51.2	52.7	34.5	1.44	2.00
60.0	80.0	90.0	120.0	51.2	55.8	12.1	1.63	2.13
60.0	80.0	90.0	125.0	51.2	58.5	8.0	1.68	2.26
60.0	80.0	95.0	120.0	51.2	52.4	43.7	1.35	1.97
60.0	80.0	95.0	125.0	51.2	55.5	13.0	1.59	2.08
60.0	80.0	100.0	125.0	51.2	52.1	60.5	1.26	1.94
60.0	85.0	90.0	120.0	54.9	55.8	60.0	1.32	1.99
60.0	85.0	90.0	125.0	54.9	58.5	16.2	1.57	2.12
65.0	80.0	90.0	110.0	45.3	49.0	13.3	1.67	2.06
65.0	80.0	90.0	115.0	45.3	52.7	7.1	1.75	2.22
65.0	80.0	90.0	120.0	45.3	55.8	5.3	1.77	2.38
65.0	80.0	90.0	125.0	45.3	58.5	4.4	1.77	2.53
65.0	80.0	95.0	115.0	45.3	48.6	14.6	1.62	2.02
65.0	80.0	95.0	120.0	45.3	52.4	7.4	1.72	2.15
65.0	80.0	95.0	125.0	45.3	55.5	5.5	1.74	2.27
65.0	80.0	100.0	120.0	45.3	48.3	16.3	1.57	1.98
65.0	80.0	100.0	125.0	45.3	52.1	7.7	1.69	2.09
65.0	85.0	90.0	115.0	50.8	52.7	27.5	1.53	2.05
65.0	85.0	90.0	120.0	50.8	55.8	11.2	1.67	2.20
65.0	85.0	90.0	125.0	50.8	58.5	7.6	1.71	2.36
65.0	85.0	95.0	120.0	50.8	52.4	33.0	1.45	2.00
65.0	85.0	95.0	125.0	50.8	55.5	11.9	1.63	2.13
65.0	85.0	100.0	125.0	50.8	52.1	41.7	1.36	1.97

Table 3.4 Derived thermodynamic design data for absorption heat pump systems operating on water/Carrol for heating

T_{EV} (°C)	T_{AB} (°C)	T_{CO} (°C)	T_{GE} (°C)	X_{AB} (% by weight)	X_{GE} (% by weight)	FR (dimension-less)	$(COP)_{EH}$ (dimension-less)	$(COP)_{CH}$ (dimension-less)
65.0	90.0	90.0	120.0	54.5	55.8	42.9	1.43	2.03
65.0	90.0	90.0	125.0	54.5	58.5	14.7	1.62	2.19
65.0	90.0	95.0	125.0	54.5	55.5	57.8	1.33	1.99
70.0	85.0	90.0	110.0	44.8	49.0	11.8	1.73	2.12
70.0	85.0	90.0	115.0	44.8	52.7	6.7	1.78	2.33
70.0	85.0	90.0	120.0	44.8	55.8	5.1	1.79	2.53
70.0	85.0	90.0	125.0	44.8	58.5	4.3	1.79	2.72
70.0	85.0	95.0	115.0	44.8	48.6	12.8	1.68	2.06
70.0	85.0	95.0	120.0	44.8	52.4	6.9	1.75	2.22
70.0	85.0	95.0	125.0	44.8	55.5	5.2	1.77	2.38
70.0	85.0	100.0	120.0	44.8	48.3	14.0	1.63	2.02
70.0	85.0	100.0	125.0	44.8	52.1	7.2	1.72	2.15
70.0	90.0	90.0	115.0	50.4	52.7	22.9	1.61	2.11
70.0	90.0	90.0	120.0	50.4	55.8	10.4	1.71	2.31
70.0	90.0	90.0	125.0	50.4	58.5	7.3	1.74	2.51
70.0	90.0	95.0	120.0	50.4	52.4	26.6	1.53	2.05
70.0	90.0	95.0	125.0	50.4	55.5	11.0	1.67	2.21
70.0	90.0	100.0	125.0	50.4	52.1	31.9	1.45	2.01
76.0	90.0	90.0	105.0	41.8	42.9	38.3	1.60	1.99
76.0	90.0	90.0	110.0	41.8	49.0	6.8	1.84	2.30
76.0	90.0	90.0	115.0	41.8	52.7	4.8	1.85	2.61
76.0	90.0	90.0	120.0	41.8	55.8	4.0	1.84	2.90
76.0	90.0	90.0	125.0	41.8	58.5	3.5	1.84	3.19
76.0	90.0	95.0	110.0	41.8	42.5	64.3	1.41	1.96
76.0	90.0	95.0	115.0	41.8	48.6	7.1	1.80	2.18
76.0	90.0	95.0	120.0	41.8	52.4	4.9	1.82	2.40
76.0	90.0	95.0	125.0	41.8	55.5	4.1	1.82	2.62
76.0	90.0	100.0	115.0	41.8	42.0	182.7	1.16	1.94
76.0	90.0	100.0	120.0	41.8	48.3	7.5	1.77	2.11
76.0	90.0	100.0	125.0	41.8	52.1	5.1	1.80	2.28

ing Carnot coefficient of performance $(COP)_{CH}$ for heating are plotted against

- the generator temperature T_{GE} in Figure 3.13;
- the flow ratio (FR) in Figure 3.14;
- the evaporator temperature T_{EV} in Figure 3.15;
- the refrigerant-poor solution concentration in the generator X_{GE} in Figure 3.16;
- the gross temperature lift $(T_{CO} - T_{EV})$ in Figure 3.17.

In addition, values for the flow ratio (FR) are plotted against the gross temperature lift $(T_{CO} - T_{EV})$ in Figure 3.18 for the water/lithium bromide and water/Carrol working pairs. These theoretical plots provide a benchmark

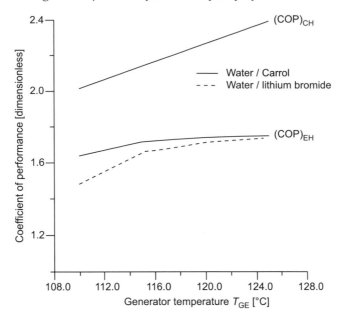

Figure 3.13 Plot of coefficients of performance $(COP)_{CH}$ and $(COP)_{EH}$ against generator temperature T_{GE} for water/lithium bromide and water/Carrol solutions.

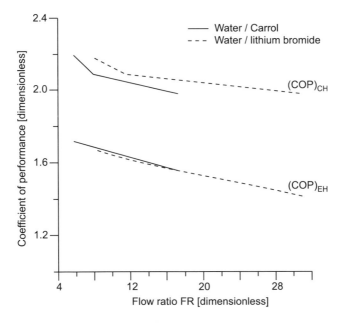

Figure 3.14 Plot of coefficients of performance $(COP)_{CH}$ and $(COP)_{EH}$ against flow ratio (FR) for water/lithium bromide and water/Carrol solutions.

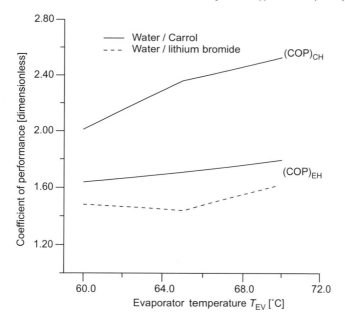

Figure 3.15 Plot of coefficients of performance $(COP)_{CH}$ and $(COP)_{EH}$ against evaporator temperature T_{EV} for water/lithium bromide and water/Carrol solutions.

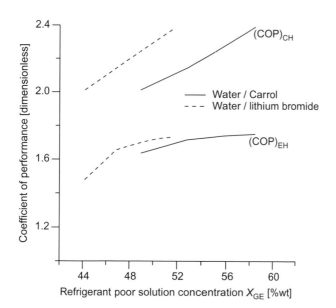

Figure 3.16 Plot of coefficients of performance $(COP)_{CH}$ and $(COP)_{EH}$ against refrigerant-poor solution concentration X_{GE} for water/lithium bromide and water/Carrol solutions.

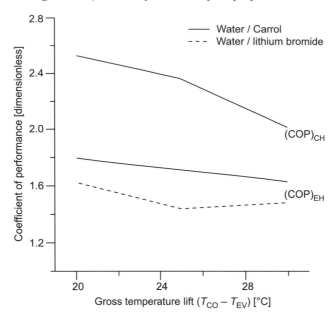

Figure 3.17 Plot of coefficients of performance $(COP)_{CH}$ and $(COP)_{EH}$ against gross temperature lift $(T_{CO} - T_{EV})$ for water/lithium bromide and water/Carrol solutions.

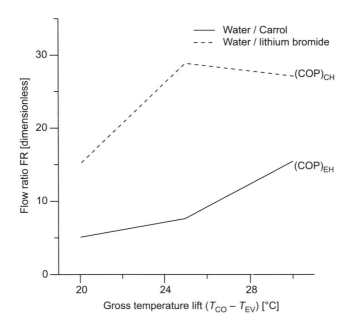

Figure 3.18 Plot of flow ratio (FR) against gross temperature lift $(T_{CO} - T_{EV})$ for water/lithium bromide and water/Carrol solutions

with which to evaluate the performance of actual absorption heat pumps including the absorption heat pump assisted effluent purification unit (AHPAEPU) which has been fabricated at the Instituto de Investigaciones Eléctricas (IIE), Cuernavaca, Mexico. This experimental unit is described in Chapter 5.

3.9 References

Best, R., Rivera, W. and Oskam, A. (1995) Thermodynamic design data for absorption heat pump systems operating on water-Carrol. Part II: Heating, *Heat Recovery Systems*, **15**(5), 435–44.

Chaudhari, S. K., Paranjape, D. V., Eisa, M. A. R. and Holland F. A., (1985) A study of the operating characteristics of a water-lithium bromide absorption heat pump, *Heat Recovery Systems*, **5**(4), 285–97.

Di Guillio, R. M., Lee, R. J., Jeter, S. M. and Teja, A. S. (1990) Properties of lithium bromide/water solutions at high temperatures and concentrations: I. Thermal conductivity, *ASHRAE Trans.*, **96**(1), 702–8.

Eisa, M. A. R., Devotta, S., Rashed, I. G. and Holland, F. A. (1986) Thermodynamic design data for absorption heat pump systems operating on water-lithium bromide. Part II: Heating, *Applied Energy*, **25**, 71–82.

Eisa, M. A. R. and Holland, F. A. (1987) A study of the optimum interaction between the working fluid and the absorbent in absorption heat pumps, *Heat Recovery Systems*, **7**(2), 107–17.

Gupta, C. P. and Sharma C. P. (1976) Entropy values of lithium bromide/water solutions and their vapours, *ASHRAE Trans.*, **82**(2), 35–46.

Khoeler, W. J., Ibele, W. E., Soltes, J. and Winter, E. R. (1987) Entropy calculations for lithium bromide/water solutions and approximation equation, *ASHRAE Trans.*, **93**, 2379–88.

Lee, R. J., Di Guillio, R. M., Jeter, S. M. and Teja, A. S. (1990) Properties of lithium bromide/water solutions at high temperatures and concentrations: II. Density and viscosity, *ASHRAE Trans.*, **96**(1), 709–28.

Macriss, R. A. and Zawaki, T. S. (1989) Absorption fluid data survey: 1989 update, *ORNL Report*, ORNL/Sub84/47989/4.

McNeely, L. A. (1979) Thermodynamic properties of aqueous solution of lithium bromide, *ASHRAE Trans.*, **85**(1), 413–34.

Moran, M. J. and Herold, K. E. (1989) Thermodynamic properties of solutions from fundamental equations of state with application to absorption heating/cooling system analysis: A review, in *Proceedings of International Symposium on Thermodynamic Analysis and Improvement of Energy Systems*, Beijing, China, June 5–8, 445–64.

Patil, K., Kim, M. N., Eisa, M. A. R. and Holland, F. A. (1989) Experimental evaluation of aqueous lithium halides as single-salt systems in absorption heat pumps, *Applied Energy*, **34**(2), 99–111.

Reimann, R. C. and Biermann, W. J. (1984) *Development of a Single Family Absorption Chiller for Use in Solar Heating and Cooling Systems, Phase III Final Report*, Prepared for the USA Department of Energy under contract EG/77/C/03/1587, Carrier Corporation, October.

4 An experimental mechanical vapour compression heat pump assisted effluent purification unit

4.1 Nomenclature

(COP)	coefficient of performance [dimensionless]
$(COP)_A$	actual coefficient of performance [dimensionless]
$(COP)_C$	Carnot coefficient of performance [dimensionless]
$(COP)_R$	Rankine coefficient of performance [dimensionless]
(CR)	compression ratio P_{CO}/P_{EV} [dimensionless]
(ER)	ratio of the isentropic to the overall efficiency of the compressor [dimensionless]
H	enthalpy per unit mass [kJ kg^{-1}]
(HPE)	heat pump effectiveness [dimensionless]
P	pressure [bar]
Q	heat flow [kW] or heat quantity [kWh]
Q_{CO}	heat flow [kW] or heat quantity [kWh] supplied to boil the brine
T	temperature [°C or K]
W	rate of work [kW] or quantity of work [kWh]
η_I	isentropic efficiency [dimensionless]

Subscripts

A	actual
AUX	auxiliary
CO	condenser
D	delivered
e	electric
EV	evaporator
R	Rankine
S	source

4.2 Introduction

In the 1980s, researchers from the Instituto de Investigaciones Eléctricas (IIE), Cuernavaca, Mexico, designed, constructed and tested several mechanical vapour compression heat pump assisted effluent purification units.

In the first stage of the work, a standard commercially available packaged heat pump was coupled to an effluent purification unit. The unit produced very good experimental results.

In the second stage of the work, the unit was redesigned to have fewer heat exchangers and an integrated heat pump. The redesigned unit produced even better results.

These heat pump assisted effluent purification units were used to produce pure distilled water from geothermal brine at the Los Azufres geothermal field in the State of Michoacan, Mexico. The quality of the distilled water was as good as that produced in the laboratory with an electrically heated distillation unit. The concentrated geothermal brine was returned to the geothermal field. Subsequently, a portable unit was fabricated at IIE, Cuernavaca, Mexico to be used as a demonstration unit.

4.3 Heat Pump Assisted Effluent Purification Unit No. 1

4.3.1 Equipment

In the first stage of the work, a model TPB20 water to water heat pump supplied by the Westinghouse Electric Co., USA, was used. The unit had two working fluid circuits designed to deliver 56 kW of heat as hot water at a temperature of 71°C from a water heat source at 52°C with a coefficient of performance (COP) of 4.74. The unit operated with R114 ($CCIF_2CCIF_2$) as the working fluid or refrigerant.

The evaporators were shell and tube heat exchangers. The R114 working fluid was on the tube side and the water on the shell side. The condensers were also shell and tube heat exchangers but, in this case, the R114 working fluid flowed through the shell and water flowed through the tubes.

The heat pump had dual compressors, expansion valves, check valves, refrigerant thermostats, flow sightglasses, isolation valves, water drains and pressure and temperature switches. Figure 4.1 is a schematic diagram of the arrangement used to evaluate the Westinghouse heat pump. A detailed description of the equipment has been given by Frias [1991] and Frias *et al.* [1991].

In order to evaluate the unit at different operating conditions and to gain operational experience, an auxiliary heat exchanger HE1 was installed to heat the pure water supplied to the evaporator EV of the heat pump. A second auxiliary heat exchanger HE2 was installed to capture the heat delivered by the heat pump condenser CO.

The instrumentation included thermocouples, pressure gauges, water flow meters, a digital voltmeter and an ammeter together with an electrical control panel. The data obtained were used to calculate the heat balances and the coefficients of performance (COP) for the heat pump.

In order to simplify the operation of the purification unit, only one of the heat pump circuits was used. Figure 4.2 is a schematic diagram of the heat

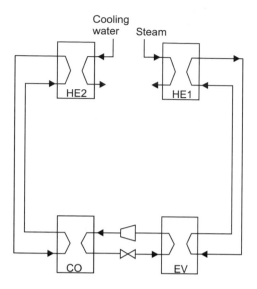

Figure 4.1 Schematic arrangement for evaluating the Westinghouse heat pump.

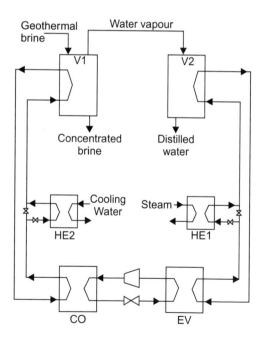

Figure 4.2 Schematic diagram of the heat pump assisted effluent purification unit No. 1.

pump brine purification unit. The evaporator and condenser vessels V1 and V2 in the purification unit had the same dimensions. The vessels were 1.99 m long and 0.34 m in diameter with a capacity of 0.120 m^3. Each had a coil 1.3 m in length with an external diameter of 0.28 m. This consisted of stainless tubing with an outside diameter of 0.019 m (0.75 in).

Pipelines of 0.0508 m (2 in) nominal diameter were used to couple the heat pump to the brine purification unit. Flanges were used to couple the coils in the vessels V1 and V2 to facilitate maintenance because of the scaling potential of geothermal brine.

Four thermocouples were installed at the inlet and outlet of the water circuits and one in the body of each of the vessels V1 and V2. These were connected to a 10-channel digital thermometer. A graduated scale water level indicator and a vacuum gauge were installed in each of the vessels V1 and V2 using oil-cloth plugs. The water level indicators were used to measure the quantity of water distilled from the geothermal brine. Vessels V1 and V2 and the pipelines were fully insulated in order to minimise heat losses.

A vacuum pump was connected to the top of vessel V2 in order to control the boiling temperature of the geothermal brine in vessel V1 in the range 60–88°C.

4.3.2 Experimental procedure

Prior to incorporating the heat pump into the purification unit shown schematically in Figure 4.2, the heat pump was evaluated separately in the test circuit shown schematically in Figure 4.1.

The crankcase compressor heaters were energised for 24 hours prior to operating the unit. The water circulation pumps were turned on and steam was supplied to the auxiliary heat exchanger HE1, shown in Figure 4.2, in order to supply heat to the evaporator. Once the selected temperatures were reached in the water circuits of the heat pump, the compressor started automatically. The heat delivered by the condenser CO of the heat pump was captured by the cooling water in the auxiliary heat exchanger HE2.

Once the preset temperatures were reached in the water circuit, the water was used to heat the geothermal brine in the evaporator V1 of the purification unit. Once the geothermal brine reached its boiling temperature, the entire water flow from the heat pump evaporator EV was sent to the condenser V2 of the purification unit. The evaporated water vapour from the boiling geothermal brine in vessel V1, then started to condense in vessel V2. In order to obtain steady-state conditions, part of the water flow from the heat pump condenser CO was diverted to the auxiliary heat exchanger HE2. In order to control the boiling temperature of the geothermal brine at a present level, a constant vacuum must be maintained in vessel V2. Steady-state conditions were assumed when the temperatures, flow rates and electric power consumption were constant for at least 30 minutes.

The temperature of the water supplied by the heat pump condenser CO to boil the geothermal brine in vessel V1, ranged from 70°C to 94°C. The temperature of the water supplied by the heat pump evaporator EV to condense the water vapour in vessel V2 ranged from 35°C to 56°C. The temperatures, pressures, flow rates and electrical consumption were continuously registered at steady-state conditions for each experimental run. Additionally, the flow rates of the water distilled from the boiling brine together with the vacuum pressure and the temperature of the boiling brine were also recorded for steady-state conditions in each experimental run.

4.3.3   *Experimental evaluation of the heat pump*

Sixteen experimental runs were carried out to evaluate the heat pump in the layout shown schematically in Figure 4.1 prior to its incorporation into the brine purification unit shown schematically in Figure 4.2. The water flow

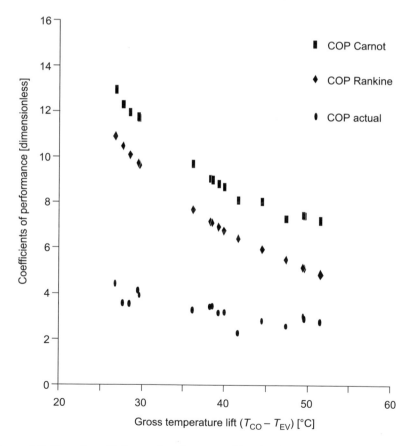

Figure 4.3 Plot of coefficients of performance $(COP)_C$, $(COP)_R$ and $(COP)_A$ against gross temperature lift $(T_{CO} - T_{EV})$ for the Westinghouse heat pump.

rates were manually controlled in order to achieve steady-state conditions. On average, steady-state conditions were reached in about 45 minutes.

Figure 4.3 is a plot of the Carnot coefficient of performance $(COP)_C$, the theoretical Rankine coefficient of performance $(COP)_R$ and the actual coefficient of performance $(COP)_A$ against the gross temperature lift $(T_{CO} - T_{EV})$ where T_{CO} and T_{EV} are the condensation and evaporation temperatures respectively of the working fluid, R114. The value of $(COP)_A$ varied from 4.5 to 2.2 for $(T_{CO} - T_{EV})$ values from 26°C to 52°C and the data followed the conventional trend [Reay and Macmichael, 1979].

The heat pump effectiveness $(HPE)_R$ defined by the equation

$$(HPE)_R = \frac{(COP)_A}{(COP)_R} \qquad (4.1)$$

is the ratio of the actual coefficient of performance $(COP)_A$ to the theoretical Rankine coefficient of performance $(COP)_R$.

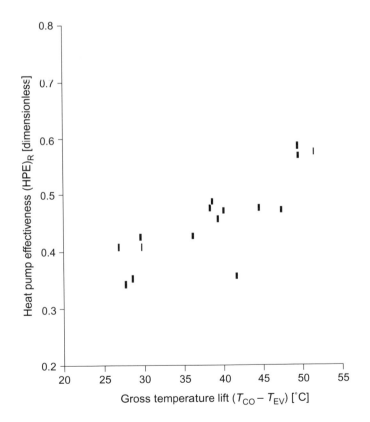

Figure 4.4 Plot of heat pump effectiveness $(HPE)_R$ against gross temperature lift $(T_{CO} - T_{EV})$ for the Westinghouse heat pump.

The plot of $(HPE)_R$ against $(T_{CO} - T_{EV})$ in Figure 4.4 shows that $(HPE)_R$ increases with an increase in $(T_{CO} - T_{EV})$ with a maximum $(HPE)_R$ value of 0.6 at $(T_{CO} - T_{EV}) = 50\,°C$. This relatively low value is because of the low isentropic and overall efficiencies of the available compressor. $(HPE)_R$ values of 0.8 or more could be obtained with a more appropriate compressor.

Figure 4.5 is a plot of the ratio of the isentropic to overall efficiency of the compressor (ER) against the compression ratio $(CR) = P_{CO}/P_{EV}$ where P_{CO} and P_{EV} are the condensing and evaporating pressures respectively of the working fluid, R114. The isentropic efficiency η_I was calculated from the following equation [Holland *et al.*, 1982]

$$\eta_I \cong 1 - 0.05(CR) \tag{4.2}$$

The highest value $(ER) = 0.95$ [Baumeister *et al.*, 1978] corresponds to a value $(CR) = 3.3$. The lowest values for (ER) correspond to the lowest (CR) values.

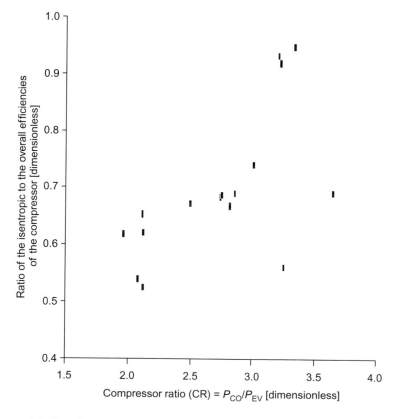

Figure 4.5 Plot of ratio (ER) of isentropic to overall efficiency of the Westinghouse compressor against compression ratio (CR).

4.3.4   *Experimental evaluation of the heat pump assisted brine purification unit No. 1*

Seventeen experimental runs were carried out to evaluate the heat pump assisted purification unit shown schematically in Figure 4.2. Batch distillation was used in the tests. Seven runs were made with water in vessel V1 and ten runs were made with geothermal brine in the same vessel. These experimental runs were carried out for different water and brine boiling temperatures between 60°C and 85°C.

The capacity of the geothermal brine purification unit is related to the quantity of heat delivered to the heat pump evaporator Q_{EV}. In a batch distillation process, the difference between the quantity of heat generated in the heat pump condenser Q_{CO} and that delivered to the heat pump evaporator Q_{EV} must be rejected by an auxiliary heat exchanger HE2 with a capacity $(Q_{CO} - Q_{EV})$ in order to achieve steady-state conditions.

In order to obtain precise control of the boiling temperature in vessel V1 of Figure 4.2, it was necessary to adjust the vacuum in vessel V2. The range of vacuum applied was between 0.17 and 0.53 bar (13 and 40 cm Hg). The condensed water temperatures in vessel V2 were between 45°C and 60°C.

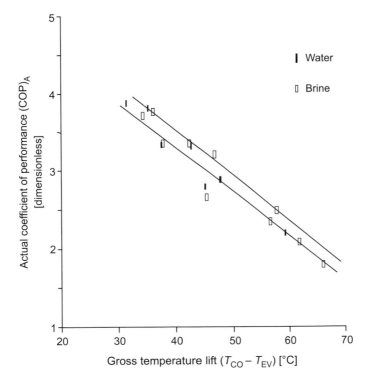

Figure 4.6 Plot of actual coefficient of performance $(COP)_A$ against gross temperature lift $(T_{CO} - T_{EV})$ for the heat pump assisted effluent purification unit No. 1.

The temperatures in vessels V1 and V2 were determined by the amount of vacuum applied to the unit.

Figure 4.6 is a plot of the actual coefficient of performance $(COP)_A$ against the gross temperature lift $(T_{CO} - T_{EV})$ for the heat pump assisted brine purification unit. The upper line corresponds to the experiments with the highest brine or water boiling temperature and the lower line to the experiments with the lowest brine or water boiling temperature.

Figure 4.7 is a plot of the theoretical Rankine coefficient of performance $(COP)_R$ against the condensing temperature T_{CO} for the working fluid

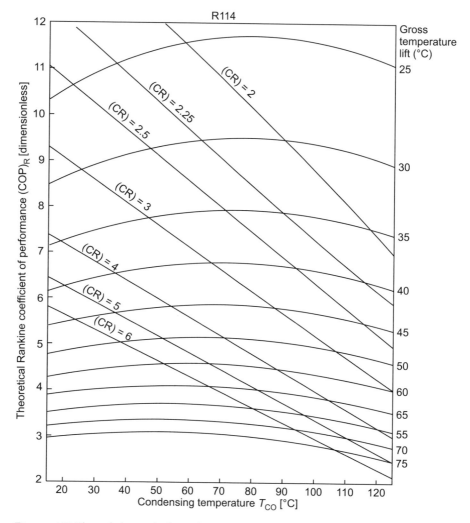

Figure 4.7 Plot of theoretical Rankine coefficient of performance $(COP)_R$ against condensation temperature T_{CO} for various values of compression ratio (CR) and gross temperature lift $(T_{CO} - T_{EV})$ for R114.

R114 at various values of the compression ratio (CR) and gross temperature lift $(T_{CO} - T_{EV})$ [Holland *et al.*, 1982]. Individual sets of data are observed in Figure 4.6. The two sets of data differ as a result of the differences in condensing temperature T_{CO} of the R114 working fluid (Figure 4.7).

Supranto *et al.* [1987] carried out experiments on a mechanical vapour compression heat pump operating with R114 as the working fluid at different values of the condensing temperature T_{CO}. These authors also observed two similar sets of data.

Figure 4.8 is a plot of $(COP)_A$ against the production rate of distilled water with both water and geothermal brine. It is seen that the production rate of distilled water increases as $(COP)_A$ increases.

Figure 4.9 is a plot of the gross temperature lift $(T_{CO} - T_{EV})$ against the production rate of distilled water which shows a decrease in the production rate of distilled water with an increase in $(T_{CO} - T_{EV})$.

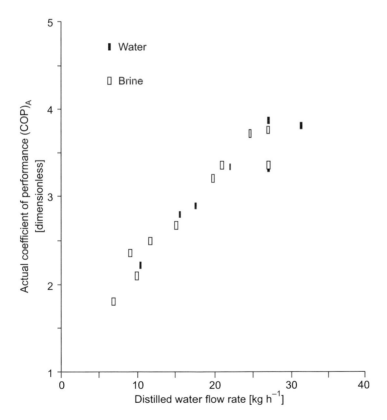

Figure 4.8 Plot of actual coefficient of performance $(COP)_A$ against distilled water flow rate for the heat pump assisted effluent purification unit No. 1.

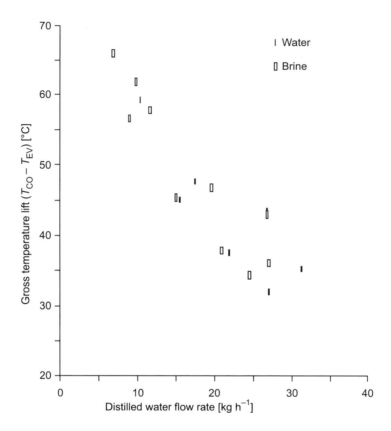

Figure 4.9 Plot of gross temperature lift ($T_{CO} - T_{EV}$) against distilled water flow rate for the heat pump assisted effluent purification unit No. 1.

The figures demonstrate that the highest production rate of distilled water is obtained for the highest value of $(COP)_A$ and the lowest value of $(T_{CO} - T_{EV})$.

Chemical analyses were carried out in the laboratory on samples taken from the geothermal brine and distilled water in the heat pump assisted brine purification unit. Colorimetric and volumetric methods were used to determine the silica and chloride concentrations. Approximately 1000 and 1300 parts per million (ppm) of silica were found in the initial and final geothermal brine samples respectively. Tables 4.1 and 4.2 show the silica and chloride concentrations for the input and output geothermal brine and also for the distilled water. The 1.0 ppm concentration for the latter was lower than the 5.5 ppm concentration obtained for salt free water distilled in a laboratory distillation unit, which demonstrates the efficiency of the experimental equipment for the heat pump assisted purification of geothermal brine.

Table 4.1 Silica concentration in brine and distilled water from the heat pump assisted purification unit No.1

	Silica concentration (ppm)		
Sample number	Inlet brine	Outlet brine	Distilled water
1	1006	1128	3.1
2	963	1120	1.4
3	935	1192	1.0
4	921	1228	1.2
5	1012	1346	1.5
6	994	1318	2.0
7	4030	6045	4.3
8	1012	1314	2.3
9	1018	1474	1.8
10	1012	1476	0.6

Table 4.2 Chloride concentration in brine and distilled water from the heat pump assisted purification unit No.1

	Chloride concentration (ppm)		
Sample number	Inlet brine	Outlet brine	Distilled water
1	4085	5925	6.2
2	4068	6038	9.8
3	4097	5986	20.6
4	4030	7340	8.7
5	4102	5901	10.2
6	4030	5829	6.3
7	4030	6045	4.3
8	4030	5829	4.3
9	4030	6477	3.6
10	4052	5792	3.4

4.4 Heat Pump Assisted Effluent Purification Unit No. 2

4.4.1 Equipment

In the second phase of the work, an integrated heat pump assisted effluent purification unit was designed, constructed and tested [Siqueiros *et al.*, 1992, 1995]. This is shown schematically in Figure 4.10. The main difference between the integrated unit and the arrangement used in the first phase of the work was the elimination of two heat exchangers and the reduced amount of tubing for the water circuits. This resulted in higher efficiencies, due to the lower temperature differences in the heat exchangers and the reduction in the number of components in the unit.

The condensing R114 working fluid supplied heat to the geothermal brine in vessel V1. Excess heat was rejected in the auxiliary condenser. The condensed R114 working fluid then flowed via a thermostatic expansion valve to vessel V2 where it evaporated, extracting a quantity of heat Q_{EV} from the condensing water vapour which had been distilled from the geothermal brine. The R114 working fluid was then compressed and returned to vessel V1 to complete the cycle. In a commercial installation, the excess heat could be rejected in an auxiliary water vapour condenser. This would increase the capacity of the unit using essentially the same total heat transfer area. This heat could be used to preheat the geothermal brine supply.

In the integrated unit shown schematically in Figure 4.10, the R114 flowed inside coils in vessels V1 and V2 which were flanged to facilitate disconnection when necessary. The compressor in Figure 4.10 was driven by a 7.44 kW (10 hp) electric motor. The heat pump circuit had a check valve, an oil separator, a solenoid valve, a sight glass, an orifice plate flow meter with a pressure transducer and a charge valve.

The vacuum pump and the 60 litre tank, that were used in the previous purification unit, were installed in the new brine purification unit as auxiliary equipment. When it was necessary to remove distilled water or brine, a slightly higher vacuum was created in the auxiliary tank to withdraw liquid from either of the vessels V1 and V2 in Figure 4.11.

For the initial heating of the contents of vessel V1, a 10 kW electric resistance heater was installed. An auxiliary heat exchanger was installed in the R114 working fluid circuit in Figure 4.10 in order to reject the excess heat $(Q_{CO} - Q_{EV})$ required to balance the system. The instrumentation included type T thermocouples, Bourdon tube pressure gauges, rotameters to measure the water flow rates, an orifice plate, a digital voltmeter and a barometer.

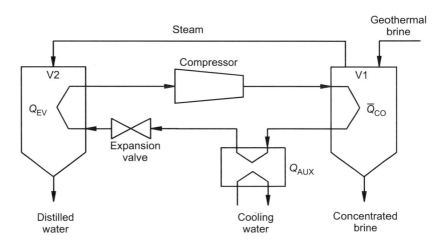

Figure 4.10 Schematic diagram of the integrated heat pump assisted effluent purification unit.

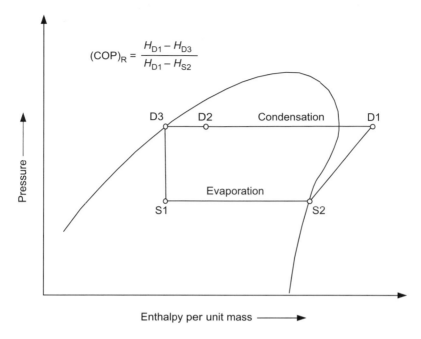

Figure 4.11 The ideal heat pump cycle in a plot of pressure *P* against enthalpy per unit mass *H*.

4.4.2   *Experimental procedure*

Prior to start up, the compressor oil in the crankcase was heated for 24 hours to evaporate the dissolved R114 working fluid. Vessel V2 in Figure 4.10 was filled to a minimum level with water which was heated to a temperature of 50 °C by the electric resistance heater. Vessel V1 in Figure 4.10 was filled with geothermal brine. The heat pump compressor was then started. When the geothermal brine reached a temperature of 60°C, the vacuum pump was started in order to control the boiling temperature of the brine by controlling the pressure in the unit. When the desired boiling temperature was reached, the electric heater was switched off. When the steady state had been achieved for one hour, the operating data were recorded during the following hour. The flow rate of the cooling water in the auxiliary condenser shown in Figure 4.10 was adjusted to reject the excess heat ($Q_{CO} - Q_{EV}$) from the working fluid circuit in order to maintain the steady state.

4.4.3   *Experimental evaluation of the heat pump assisted brine purification unit No. 2*

Twenty-three experimental runs were carried out to evaluate the heat pump assisted purification unit shown schematically in Figure 4.10. The actual coefficient of performance $(COP)_A$ obtained ranged from 2.5 to 4.5 and the production rates of distilled water ranged from 14.4 to 37.8 kg h^{-1}. The distilled water produced was equivalent to commercially available distilled water with respect to its silica and chloride levels.

The capacity of the heat pump assisted purification unit shown schematically in Figure 4.10 is limited by the heat pump evaporator duty Q_{EV}. In order to obtain steady-state operation, the auxiliary heat exchanger must reject a quantity of heat equivalent to the work done by the compressor. An energy balance on a mechanical vapour compression heat pump is given by the equation

$$Q_{CO} = Q_{EV} + W \tag{4.3}$$

where Q_{CO} and Q_{EV} are the duties of the heat pump condenser and evaporator respectively and W is the work input from the compressor.

For the heat pump assisted purification system, the energy balance needs to be modified to take into account the heat rejected Q_{AUX} in the auxiliary heat exchanger given by the equations

$$Q_{AUX} = Q_{CO} - Q_{EV} \tag{4.4}$$

and

$$\bar{Q}_{CO} + Q_{AUX} = Q_{EV} + W \tag{4.5}$$

where

$$\bar{Q}_{CO} = Q_{EV} \tag{4.6}$$

The energy balance can be written in terms of enthalpies by the equation

$$(H_{D1} - H_{D2}) + (H_{D2} - H_{D3}) = (H_{S2} - H_{S1}) + (H_{D1} - H_{S2}) \tag{4.7}$$

where the state points are shown in Figure 4.11 which is a plot of the pressure P of the working fluid against the enthalpy per unit mass H for a mechanical vapour compression heat pump.

The experimentally determined actual coefficient of performance $(COP)_A$ was calculated from the equation

$$(COP)_A = \frac{(\bar{Q}_{CO} + Q_{AUX})}{W_e} \tag{4.8}$$

where W_e is the electric power supplied to the compressor.

Figure 4.12 is a plot of the Carnot, the theoretical Rankine and the experimentally determined actual coefficients of performance $(COP)_C$,

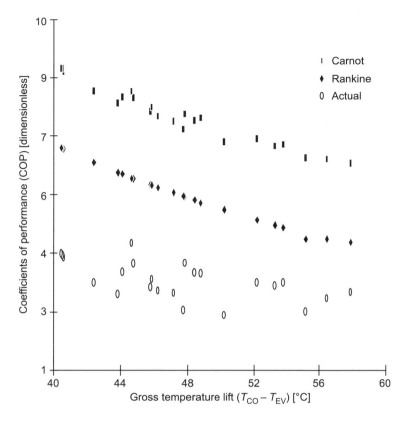

Figure 4.12 Plot of coefficients of performance $(COP)_C$, $(COP)_R$ and $(COP)_A$ against gross temperature lift $(T_{CO} - T_{EV})$ for the heat pump assisted effluent purification unit No. 2.

$(COP)_R$ and $(COP)_A$, respectively against the gross temperature lift $(T_{CO} - T_{EV})$ where T_{CO} and T_{EV} are the condensing and evaporating temperatures respectively of the R114 working fluid.

Figure 4.13 is a plot of $(COP)_A$ against $(T_{CO} - T_{EV})$ on a larger scale. The scatter of the data is due to the variation in operating conditions during the test runs. The main variation in operating conditions was the condensing temperature T_{CO} and hence the condensing pressure P_{CO}. The experimentally determined actual coefficient of performance $(COP)_A$ decreases as the gross temperature lift $(T_{CO} - T_{EV})$ increases in the same way as recorded by previous workers [Frias, 1991; Supranto *et al.*, 1987].

Figure 4.14 is a plot of the rate of heat extracted Q_{EV} in vessel V2 against the rate of heat supplied $\bar{Q}_{CO}$ to vessel V1 for the heat pump assisted purification unit shown schematically in Figure 4.10. There is good agreement between the two heat rates with the differences due to the small heat losses in the unit.

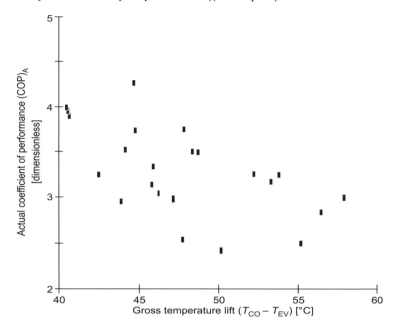

Figure 4.13 Plot of coefficient of performance $(COP)_A$ against gross temperature lift $(T_{CO} - T_{EV})$ for the heat pump assisted effluent purification unit No. 2.

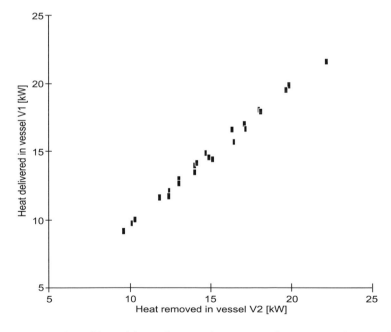

Figure 4.14 Plot of heat delivered in vessel V1 against heat removed in vessel V2 for the heat pump assisted effluent purification unit No. 2.

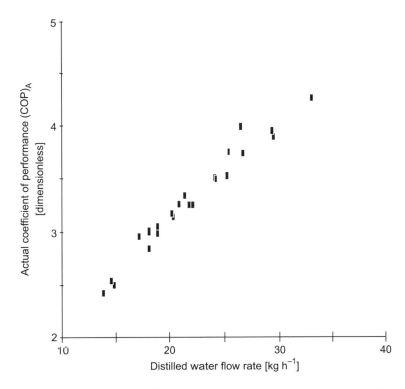

Figure 4.15 Plot of actual coefficient of performance $(COP)_A$ against distilled water flow rate for the heat pump assisted effluent purification unit No. 2.

Figure 4.15 is a plot of the experimentally determined actual coefficient of performance $(COP)_A$ against the production rate of distilled water for the unit.

Figure 4.16 is a plot of the heat pump effectiveness $(HPE)_R$ against the gross temperature lift $(T_{CO} - T_{EV})$ where $(HPE)_R$ is defined by the equation

$$(HPE)_R = \frac{(COP)_A}{(COP)_R} \tag{4.1}$$

The $(HPE)_R$ varies from 0.48 to 0.72 with a tendency to increase as the gross temperature lift $(T_{CO} - T_{EV})$ also increases.

Chemical analyses were carried out to determine the concentrations of silica, chloride, boron, sulphate, sodium, potassium, lithium, calcium and arsenic in the distilled water and the input and concentrated output geothermal brine. The results of these analyses are given in Tables 4.3–4.5. The silica concentration in the input geothermal brine was 1000 ppm and in the concentrated output geothermal brine the concentrations ranged between 1246 and 1621 ppm. The highest silica concentration in distilled water was

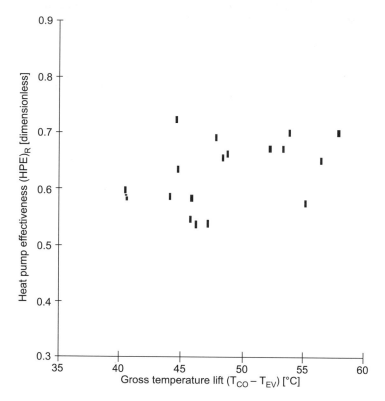

Figure 4.16 Plot of heat pump effectiveness (HPE)$_R$ against gross temperature lift ($T_{CO} - T_{EV}$) for the heat pump assisted effluent purification unit No. 2.

2.4 ppm but most of the results were less than half this value and were lower than previously obtained by Frias [1991]. There was a similar reduction in the concentrations of other substances in the distilled water which had a quality comparable to that obtained from distilling tap water in a laboratory still. The concentrations of six substances in the distilled water were so low that they could not be measured. The concentrated geothermal brine could perhaps be used to obtain byproducts such as lithium [Angulo *et al.*, 1987].

It can be concluded that the experimental integrated heat pump assisted purification unit has worked very well. The quality of the distilled water obtained from the geothermal brine was similar to that of commercially available distilled water with respect to chlorides and silica.

A distilled water production rate of 37.8 kg h^{-1} with an actual coefficient of performance $(COP)_A = 4.5$ and a brine boiling temperature of 63°C was achieved. The maximum thermal capacity of the unit was 24.5 kW$_t$.

If the compressor had been better matched to the system, $(COP)_A$ values of 5.25 or better could have been obtained and heat pump effectiveness (HPE)$_R$

Table 4.3 Results of the chemical analyses of the brine supplied to the heat pump assisted purification unit No. 2 (ppm)

Test number	Silica	Chlorides	Boron	Sulphates	Sodium	Potassium	Lithium	Calcium	Arsenic
1	989	4085	270.5	14.8	1892	486	22.5	16.1	32.2
2	985	4023	272.4	14.9	1910	489	21.6	16.4	32.6
3	1002	4112	270.5	15.3	1902	495	23.4	16.8	33.0
4	998	4184	271.0	14.7	1874	492	23.1	15.9	11.8
5	1008	4182	270.5	16.0	1895	496	22.9	16.0	31.6
6	1010	4184	275.0	15.8	1912	500	24.0	16.3	32.4
7	1009	4185	281.3	15.7	1921	488	23.2	16.7	32.8
8	997	4056	281.0	14.8	1878	495	23.0	16.8	31.4
9	1015	4186	270.5	14.6	1915	487	21.9	16.2	33.1
10	1017	4187	276.5	14.9	1918	482	22.1	16.0	33.4
11	1010	4184	270.5	15.1	1923	491	22.8	15.9	33.6
12	1008	4186	281.3	15.6	1889	490	22.6	15.8	32.7
13	1010	4184	270.5	15.7	1892	496	23.4	14.9	32.5
14	1012	4182	272.5	15.8	1901	501	23.1	16.2	31.9
15	1011	4180	281.3	14.9	1912	487	22.7	16.4	29.8
16	1045	4188	276.5	16.0	1894	492	22.9	15.7	31.2
17	1091	4149	281.3	16.1	1869	496	21.9	15.6	31.0
18	1006	4175	281.3	15.8	1875	493	22.0	15.8	33.1
19	1021	4182	278.3	15.7	1913	487	22.5	15.9	32.4
20	1006	4184	270.5	15.4	1932	483	23.4	16.1	32.7
21	1007	4126	270.5	15.3	1918	500	23.1	15.4	32.1
22	1006	4114	270.5	14.9	1919	501	22.8	15.2	32.8
23	1012	4115	281.3	14.7	1889	487	22.4	15.7	32.4
24	1009	4139	270.5	15.3	1893	497	21.9	14.9	32.0
25	1016	4187	281.3	16.0	1910	491	21.6	15.6	31.9
26	1005	4173	274.3	16.1	1914	493	22.4	15.5	33.1
27	1003	4184	281.3	15.9	1920	491	23.1	15.3	33.0
28	1007	4088	276.5	15.2	1889	501	23.5	15.1	32.6
29	1001	4110	281.3	15.0	1945	481	21.9	15.0	32.4
30	985	4083	270.5	15.4	1895	480	22.8	16.1	32.9
31	1005	4171	281.3	14.7	1915	486	22.9	16.0	32.0
32	994	4148	274.6	14.9	1916	491	23.2	15.9	32.1
33	984	4056	270.5	15.3	1925	497	23.5	15.4	31.8
34	956	4110	281.3	15.6	1875	482	22.8	15.2	29.9
35	991	4122	270.5	16.0	1886	493	22.6	16.0	31.5
36	979	4149	281.3	15.8	1913	485	23.1	15.4	11.1
Average	1006	4146	275.6	15.8	1904	491	22.7	15.8	32.2

Table 4.4 Results of the chemical analyses of the concentrated brine from the heat pump assisted purification unit No. 2 (ppm)

Test number	Silica	Chlorides	Boron	Sulphates	Sodium	Potassium	Lithium	Calcium	Arsenic
1	1345	5584	371.6	19.8	2569	657.7	30.5	21.4	44.5
2	1428	5973	389.5	21.3	2858	721.4	32.3	23.2	46.9
3	1621	6875	421.9	24.9	3204	839.4	37.5	26.8	49.4
4	1615	6986	421.5	24.8	3146	832.1	36.8	25.1	48.3
5	1589	6991	421.9	26.0	3195	827.1	37.5	26.3	49.8
6	1635	6892	435.6	27.0	3272	843.5	38.6	26.4	49.2
7	1468	6075	421.9	23.5	2801	721.2	33.4	23.4	47.6
8	1305	5278	389.5	18.8	2459	653.1	30.4	21.2	42.5

Table 4.4 (contd.)

Test number	Silica	Chlorides	Boron	Sulphates	Sodium	Potassium	Lithium	Calcium	Arsenic
9	1349	5593	411.2	19.4	2585	650.7	29.7	21.8	44.3
10	1482	6125	398.2	21.2	2854	715.8	31.9	23.4	48.6
11	1433	6287	424.9	21.9	2909	740.5	33.9	23.4	49.9
12	1387	5897	398.5	22.0	2657	684.9	32.1	22.2	46.4
13	1455	6204	411.2	23.5	2825	740.5	34.4	21.6	49.2
14	1485	6281	398.5	23.4	2844	751.8	33.9	23.7	48.3
15	1415	6009	400.3	21.8	2818	712.5	33.5	23.3	43.7
16	1509	6380	411.2	23.5	2978	742.8	34.6	24.2	47.6
17	1416	5476	389.5	21.1	2478	645.6	29.2	19.8	42.1
18	1502	6344	411.2	23.4	2959	745.3	33.7	23.7	49.9
19	1465	6123	389.5	23.1	2800	718.5	33.1	23.1	48.1
20	1386	5822	389.5	21.1	2684	685.1	32.4	21.6	45.6
21	1344	5612	389.5	20.2	2604	684.2	31.1	20.8	44.0
22	1313	5500	389.5	19.5	2546	668.4	30.5	20.6	43.7
23	1484	6054	411.2	21.8	2812	723.6	33.2	23.1	47.9
24	1411	5739	389.5	21.3	2623	689.4	31.1	20.1	45.1
25	1365	5620	378.7	21.2	2586	652.1	29.8	19.8	44.2
26	1312	5271	350.2	20.0	2512	635.4	28.8	19.4	42.6
27	1542	6463	378.7	23.1	2989	760.5	34.9	23.3	50.0
28	1535	6288	411.2	23.2	2945	768.8	36.2	22.9	49.2
29	1433	5846	411.2	21.2	2748	682.8	32.0	21.0	46.2
30	1297	5512	389.5	19.9	2580	648.2	30.4	20.9	44.6
31	1325	5472	389.5	19.1	2504	632.7	29.9	20.8	42.8
32	1466	6148	389.5	21.9	2821	724.6	34.1	22.7	48.2
33	1371	5676	378.7	21.3	2685	694.1	32.6	21.1	44.8
34	1244	5413	378.7	19.8	2500	631.7	29.8	19.8	40.1
35	1246	5184	378.7	19.9	2356	620.5	29.0	19.4	39.2
36	1264	5786	411.2	21.8	2695	678.8	31.8	20.5	43.2

Table 4.5 Results of the chemical analyses of the condensate from the heat pump assisted purification unit No.2 (ppm)

Test number	Silica	Chlorides	Boron	Sulphates	Sodium	Potassium	Lithium	Calcium	Arsenic
1	< 1	3.0	< 1	N.D.	N.D.	N.D.	N.D.	N.D.	N.D.
2	< 1	2.8	< 1	N.D.	N.D.	N.D.	N.D.	N.D.	N.D.
3	< 1	3.1	1.1	N.D.	N.D.	N.D.	N.D.	N.D.	N.D.
4	< 1	4.2	1.1	N.D.	N.D.	N.D.	N.D.	N.D.	N.D.
5	< 1	4.9	1.1	N.D.	N.D.	N.D.	N.D.	N.D.	N.D.
6	< 1	4.0	1.1	N.D.	N.D.	N.D.	N.D.	N.D.	N.D.
7	< 1	4.9	1.1	N.D.	N.D.	N.D.	N.D.	N.D.	N.D.
8	1.8	5.2	1.2	N.D.	N.D.	N.D.	N.D.	N.D.	N.D.
9	1.1	5.3	1.1	N.D.	N.D.	N.D.	N.D.	N.D.	N.D.
10	< 1	3.9	1.3	N.D.	N.D.	N.D.	N.D.	N.D.	N.D.
11	< 1	5.3	1.7	N.D.	N.D.	N.D.	N.D.	N.D.	N.D.
12	< 1	4.2	1.4	N.D.	N.D.	N.D.	N.D.	N.D.	N.D.
13	< 1	6.0	1.7	N.D.	N.D.	N.D.	N.D.	N.D.	N.D.
14	< 1	2.5	< 1	N.D.	N.D.	N.D.	N.D.	N.D.	N.D.
15	< 1	6.3	1.1	N.D.	N.D.	N.D.	N.D.	N.D.	N.D.
16	< 1	6.7	1.2	N.D.	N.D.	N.D.	N.D.	N.D.	N.D.
17	1.4	5.4	2.1	N.D.	N.D.	N.D.	N.D.	N.D.	N.D.
18	1.1	4.3	1.3	N.D.	N.D.	N.D.	N.D.	N.D.	N.D.

Test number	Silica	Chlorides	Boron	Sulphates	Sodium	Potassium	Lithium	Calcium	Arsenic
19	< 1	4.6	1.2	N.D.	N.D.	N.D.	N.D.	N.D.	N.D.
20	< 1	4.6	1.1	N.D.	N.D.	N.D.	N.D.	N.D.	N.D.
21	< 1	3.2	< 1	N.D.	N.D.	N.D.	N.D.	N.D.	N.D.
22	< 1	3.9	1.1	N.D.	N.D.	N.D.	N.D.	N.D.	N.D.
23	< 1	3.2	1.1	N.D.	N.D.	N.D.	N.D.	N.D.	N.D.
24	< 1	5.6	1.6	N.D.	N.D.	N.D.	N.D.	N.D.	N.D.
25	< 1	5.3	2.1	N.D.	N.D.	N.D.	N.D.	N.D.	N.D.
26	1.4	5.3	2.2	N.D.	N.D.	N.D.	N.D.	N.D.	N.D.
27	1.7	5.6	3.2	N.D.	N.D.	N.D.	N.D.	N.D.	N.D.
28	1.2	4.4	< 1	N.D.	N.D.	N.D.	N.D.	N.D.	N.D.
29	1.0	4.9	1.1	N.D.	N.D.	N.D.	N.D.	N.D.	N.D.
30	1.5	3.2	1.1	N.D.	N.D.	N.D.	N.D.	N.D.	N.D.
31	1.1	4.6	1.2	N.D.	N.D.	N.D.	N.D.	N.D.	N.D.
32	1.1	9.5	1.2	N.D.	N.D.	N.D.	N.D.	N.D.	N.D.
33	1.7	5.3	2.1	N.D.	N.D.	N.D.	N.D.	N.D.	N.D.
34	1.4	4.7	1.5	N.D.	N.D.	N.D.	N.D.	N.D.	N.D.
35	1.4	5.3	2.1	N.D.	N.D.	N.D.	N.D.	N.D.	N.D.
36	2.4	8.1	2.1	N.D.	N.D.	N.D.	N.D.	N.D.	N.D.

N.D. = not detected

values of at least 0.8 instead of the 0.48 to 0.72 obtained with the available compressor.

4.5 References

Angulo, R., Gonzáles, J. and Lam, L. (1987) Developments in geothermal energy in Mexico - Part 10: Solar evaporation applied to chemical recovery from Cerro Prieto I brines. *Heat Recovery Systems and CHP*, 7(2), 129–38.

Baumeister, T., Avallone, E. A. and Baumeister III, T. (1978) *Mark's Standard Hand-Book for Mechanical Engineers*, Chap. 14, McGraw-Hill, New York, USA, pp. 31–6.

Frias, J. L. (1991) An experimental study of a heat pump assisted purification system for geothermal brine, MSc Thesis, University of Salford, UK.

Frias, J. L., Siqueiros, J., Fernández, H., García, A. and Holland, F. A. (1991) Developments in geothermal energy in Mexico – Part 36. The commissioning of a heat pump assisted brine purification system, *Heat Recovery Systems and CHP*, 11(4), 297–310.

Holland, F. A., Watson, F. A. and Devotta, S. (1982) *Thermodynamic Design Data for Heat Pump Systems*, Pergamon Press, Oxford, UK.

Reay, D. A. and Macmichael, D. B. A. (1979) *Heat Pumps Design and Application*, Pergamon Press, Oxford, UK.

Siqueiros, J., Fernández, H., Heard, C. and Barragán, D. (1992) *Desarrollo e Implantación de Tecnologia de Bombas de Calor, Final Report: INFORME IIE/FE/1 1/2963/F*, Mexico.

Siqueiros, J., Heard, C. L. and Holland, F. A. (1995) The commissioning of an integrated heat pump-assisted geothermal brine purification system, *Heat Recovery Systems and CHP*, 15(7), 655–64.

Supranto, S., Jaganathan, R., Dodda, S., Diggory, P. J. and Holland, F. A. (1987) Heat pump assisted distillation. IV: An experimental comparison of R114 and R11 as the working fluid in external heat pump, *Energy Research* 11, 21–33.

5 An experimental absorption heat pump assisted effluent purification unit

5.1 Nomenclature

(COP)	coefficient of performance [dimensionless]
FR	flow ratio [dimensionless]
f_L	Lang factor [dimensionless]
H	enthalpy per unit mass [kJ kg^{-1}]
HPE	heat pump effectiveness [dimensionless]
M	mass flow rate [kg s^{-1}]
P	pressure [kPa]
Q	thermal load [kW]
T	temperature [°C or K]
X	concentration of salt by weight [%]

Subscripts

A	actual
AB	absorber
CO	condenser
EH	enthalpic
EV	evaporator
GE	generator
H	heating
ir	refractive index
R	refrigerant
r	reading

5.2 Introduction

This chapter describes the design and construction of an experimental absorption heat pump assisted effluent purification unit. First, the design considerations are discussed. Then, the equations related to this design are outlined. Next, a detailed description of the design and characteristics of the main components of this experimental unit are presented. Finally, the auxiliary systems used in this experimental unit are described.

5.3 Design Considerations

An absorption heat pump assisted effluent purification unit (AHPAEPU) was designed, constructed, installed and operated at the Geothermal Department of the Instituto de Investigaciones Eléctricas (IIE), Cuernavaca, Mexico.

A simplified block diagram of the absorption heat pump assisted effluent purification system is shown in Figure 5.1. This diagram illustrates the energy flows involved in the effluent purification process with the respective products. This project is part of the energy research programme established at IIE for the development and application of heat recovery systems in Mexico [Siqueiros *et al.*, 1992]. It was also intended that the AHPAEPU should be used as a demonstration unit to help to convince Mexican industrialists of the value of using and re-using low grade heat in industrial processes.

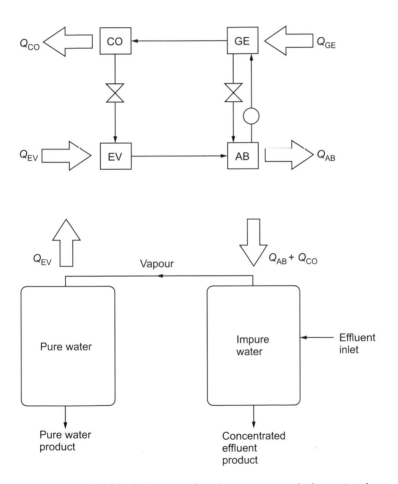

Figure 5.1 Simplified block diagram for the experimental absorption heat pump assisted effluent purification unit.

Furthermore, the AHPAEPU was designed to provide the basis for scale up to an industrial demonstration unit for evaluation over a longer period.

The AHPAEPU was designed on the basis of experience gained from previous experimental work at the IIE on electrically driven mechanical vapour compression heat pump units [Frias, 1991; Frias *et al.*, 1991; Siqueiros *et al.*, 1995]. The AHPAEPU was entirely constructed in IIE workshops using the materials available in the Geothermal Department.

This unit was designed to operate with a variety of working fluids under a wide range of operating conditions. Thus a detailed thermal design for a single working mixture under a single set of operating conditions was not justified. It was decided to design a unit which would be:

- able to operate both at vacuum or at higher than atmospheric pressure;
- simple and compact so that it could be transported to any industrial site;
- easy to fabricate and replace any of its components in case of drastic failure caused by corrosion problems;
- low cost.

5.4   Design and Characteristics of the Main Components

The main components for the AHPAEPU shown in Figure 5.2 are the generator, the absorber, the condenser, the evaporator, the auxiliary condenser and the economiser. The heat transfer area for each of the heat exchange components in the AHPAEPU were specified by Santoyo-Gutierrez (1997). These heat exchange components consisted mainly of a shell, a flange and an internal coil. Each shell was constructed from carbon steel. Each flange was designed with a small slot to contain an O-ring to seal the vessel. The O-rings were made from 4.89 mm diameter Buna-N from the Parker Co., USA, and were 938.2 mm long. Each coil was constructed from 19.05 mm diameter 316 stainless steel tubing and had 13.4, 150 mm diameter turns and a total length of 6680 mm. The pitch between each turn was 19.05 mm (Figure 5.3).

The sprayers used in the generator and absorber components were constructed from 12.7 mm diameter 316 stainless steel tubing. Each sprayer was constructed with one 150 mm diameter turn to conveniently fit the coils in the generator and the absorber, respectively. Both sprayers were perforated using a 1.58 mm diameter drill bit. The pitch between the perforations was 2 mm. All the connections used in the AHPAEPU (i.e. tubes, elbows, couples, tees, and valves) were made from 12.7 mm and 25.4 mm diameter (schedule 40) carbon steel.

5.4.1   Generator

A detailed diagram of the generator or desorber is shown in Figure 5.4. The main function of the generator was to produce pure refrigerant vapour from

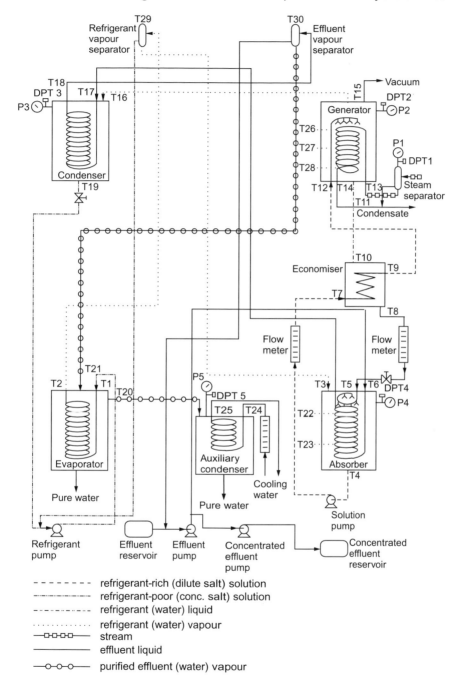

Figure 5.2 Schematic diagram of the experimental absorption heat pump assisted effluent purification unit.

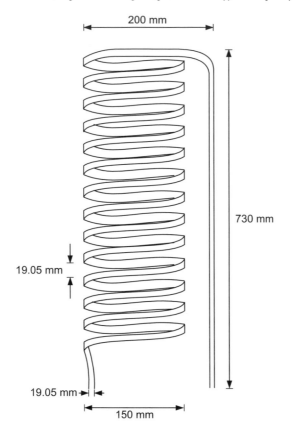

Figure 5.3 Dimensions of the stainless steel coil used in the main components of the experimental absorption heat pump assisted effluent purification unit.

the refrigerant-rich solution using the heat from a boiler or any other heat source. The generator consisted of a shell, a flange, a sprayer and a coil. There were four pipe couplings in the flange. One coupling was used to connect the generator with the economiser. The second coupling was used to mount the sprayer pipe. The last two couplings were used to secure the coil to the flange. The flange was located at the bottom of the generator.

Four perforations were made in the shell. Two of them were used to mount a vacuum gauge and a differential pressure transducer. The two remaining perforations were used to mount a glass level indicator.

In this design, the steam from the boiler was fed into the coil. The unit was designed to avoid vapour locks in the coil and the steam condensate was disposed of through a steam trap. The refrigerant-rich solution from the absorber was circulated through the sprayer located in the upper part of the coil. It was then sprayed over the coil absorbing the heat from the steam. Thus, part of the refrigerant was evaporated leaving behind a refrigerant-poor

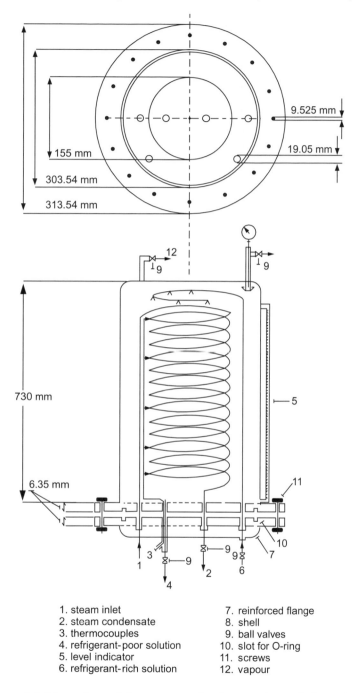

1. steam inlet
2. steam condensate
3. thermocouples
4. refrigerant-poor solution
5. level indicator
6. refrigerant-rich solution

7. reinforced flange
8. shell
9. ball valves
10. slot for O-ring
11. screws
12. vapour

Figure 5.4 Dimensions of the generator for the experimental absorption heat pump assisted effluent purification unit.

solution. This refrigerant-poor solution was then transferred back to the absorber through the economiser.

5.4.2   Condenser

A detailed diagram of the condenser is shown in Figure 5.5. The condenser was used to condense the pure refrigerant vapour from the generator. The heat of condensation was added to the heat generated in the absorber and transferred to the effluent flow from the absorber. The condenser was a shell and coil exchanger consisting of a shell, a flange and a coil. Four pipe couplings were located on the flange. Two couplings were used to secure the coil inside the condenser. The third coupling was used to connect the condenser to the generator. The fourth coupling was used to mount a vacuum gauge. The flange was mounted at the top of the condenser. Two perforations were made in the shell. The first one was connected to a glass level indicator. The second one, in the bottom of the condenser, was connected to the evaporator.

The pure refrigerant vapour from the generator was condensed on the outside surface of the coil in the condenser, as the effluent from the absorber flowed through the coil. The condensed refrigerant was then passed through a pressure reducing valve prior to entering the evaporator.

5.4.3   Evaporator

The evaporator was used to vaporise the condensed refrigerant, as the effluent vapour from the absorber is being condensed. The evaporator was a shell and coil exchanger consisting of a shell, a flange and a coil (similar to the condenser design shown in Figure 5.5). Four pipe couplings were located on the flange. Two couplings were used to secure the coil inside the evaporator. The third coupling was used to connect the evaporator to the absorber. The fourth coupling was used to mount a vacuum gauge. The flange was mounted at the top of the evaporator. Three perforations were made in the shell. The first two were connected to a glass level indicator. The third one, in the bottom of the evaporator, was connected to the absorber.

The condensed refrigerant from the condenser was evaporated inside the coil in the evaporator. The refrigerant vapour was then circulated to the absorber, as the effluent vapour was condensed on the outside surface of the coil.

In order to limit the number of perforations on the flange, two of the four drill holes had a dual function. For example, the thermocouple wiring was introduced through the same hole used to circulate the water vapour.

5.4.4   Absorber

The absorber was used to generate heat by absorbing the pure refrigerant vapour from the evaporator into the refrigerant-poor solution from the

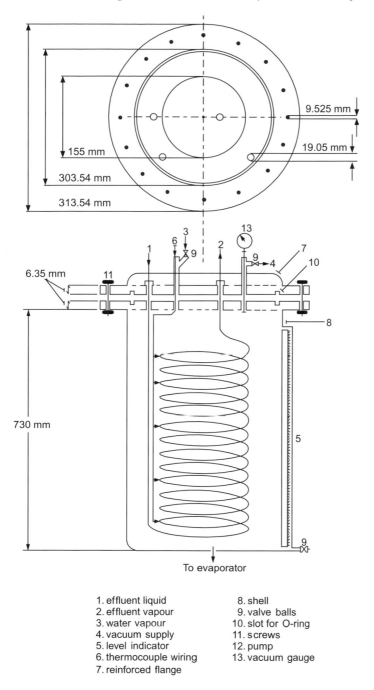

Figure 5.5 Dimensions of the condenser for the experimental absorption heat pump assisted effluent purification unit.

generator. The absorber was a shell and coil unit, a flange, a sprayer and a coil (similar to the generator design shown in Figure 5.4). Five pipe couplings were located on the flange. Two couplings were used to secure the coil inside the absorber. The third coupling was used to connect the absorber to the evaporator. The fourth coupling was used to connect the absorber to the generator. The fifth coupling was used to mount a vacuum gauge. The flange was mounted at the top of the absorber. Three perforations were made in the shell. The first two were connected to a glass level indicator. The third perforation, in the bottom of the absorber, was connected to the generator. The refrigerant vapour from the evaporator was absorbed by the refrigerant-poor solution from the generator. This solution was sprayed over the coil surface transferring the heat to the effluent that was circulated inside the coil. The refrigerant-rich solution was then pumped to the generator.

5.4.5   Economiser

In the economiser, shown in Figure 5.6, the refrigerant-rich solution coming from the absorber was pre-heated by the refrigerant-poor solution coming from the generator. The economiser which was made from 316 stainless steel consisted of a shell and a spiral coil. This was a 44C/10/34127 Model from the Heliflow Co., USA. There were four pipe couplings in the flange. Two couplings were used to connect the economiser to the generator. The other two couplings were used to connect the economiser to the absorber. The spiral coil was formed by four pipes stacked one above the other. Each pipe had a 12.7 mm diameter and a total length of 1140 mm. The hotter refrigerant-poor solution from the generator was fed through the shell. It was used to heat the cooler refrigerant-rich solution from the absorber, fed through the coil. This heat transfer reduced the heat load in the generator.

5.4.6   Auxiliary condenser

The auxiliary condenser shown in Figure 5.7 was used to reject the excess heat in the system. The auxiliary condenser consisted of a shell, a flange and a coil of tubing. Five pipe couplings were located on the flange. Two of the couplings used to secure the coil. The third coupling was used to mount a vacuum gauge. The fourth coupling was used to connect the auxiliary condenser to the evaporator. The fourth coupling was used to apply the vacuum to this part of the system. The effluent vapour from the evaporator was condensed on the outside surface of the coil by the water from the cooling tower circulating inside the coil.

5.4.7   Vapour separators

In order to separate the water and vapour, three vapour separators were designed and constructed (Figure 5.8). The design was based on a Webre separator used in the geothermal plants of Mexico and New Zealand.

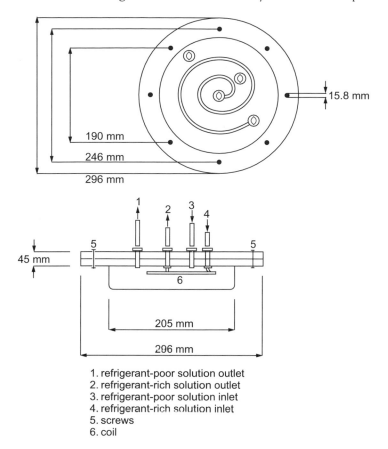

1. refrigerant-poor solution outlet
2. refrigerant-rich solution outlet
3. refrigerant-poor solution inlet
4. refrigerant-rich solution inlet
5. screws
6. coil

Figure 5.6 Dimensions of the economiser for the experimental absorption heat pump assisted effluent purification unit.

A maximum effectiveness of vapour separation for this type of separator was reported as 99.9% [Kestin *et al.*, 1980]. Separators were mounted in both the effluent circuit and the refrigerant circuits. The third separator was mounted in the steam line to remove the condensate from the generator.

The separators used in both the effluent and the refrigerant circuits failed to separate the effluent liquid properly from the purified effluent vapour. Slugs of effluent entered the evaporator adversely affecting the quality and quantity of the purified effluent liquid condensed in the unit. A similar failure occurred with the separator mounted in the refrigerant circuit. Slugs of refrigerant entered the absorber. In order to solve this problem, special liquid traps were designed and constructed. These traps included a tempered sight glass. This design allowed only purified effluent vapour and refrigerant vapour to enter the evaporator and the absorber, respectively. These arrangements significantly improved the operation of both the evaporator and the absorber.

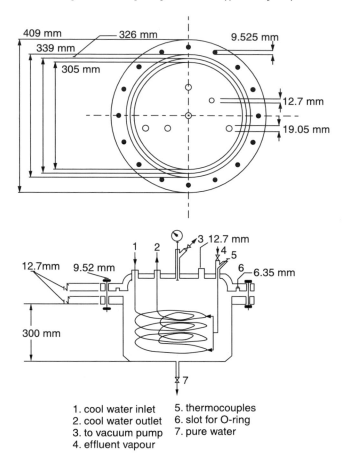

1. cool water inlet   5. thermocouples
2. cool water outlet   6. slot for O-ring
3. to vacuum pump   7. pure water
4. effluent vapour

Figure 5.7 Dimensions of the auxiliary condenser for the experimental absorption heat pump assisted effluent purification unit.

5.5   Auxiliary Equipment

The heat supply to the generator was provided by a Model 3-100-2 Ebcor Electrode Steam Boiler from the Kewanee Boiler Co., USA. This steam boiler had a $30\,kW_e$ capacity [Kewanee, 1984] capable of delivering steam at $47\,kg\,h^{-1}$, at a pressure of 6.9 bar and at a temperature of 164°C.

A cooling tower was employed to provide cooling water to the auxiliary condenser. It had a capacity of $0.75\,m^3$ and consisted of a plastic water tank divided into two sections, one filled with plastic packing and an electric fan.

In order to provide the vacuum working conditions in the AHPAEPU two vacuum pumps, a manifold and liquid traps were used.

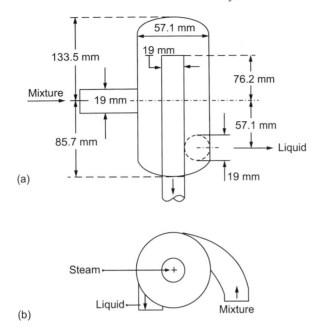

Figure 5.8 Dimensions of the Webre type separator used in the effluent and refrigerant circuits in the absorption heat pump assisted effluent purification unit: (a) longitudinal view; (b) sectional view.

5.5.1 Pumps

Rotary sliding vane type pumps were installed in the AHPAEPU which were manufactured by the Procon Products Co., USA. Rotary pumps pump fluid by means of the rotating elements which are located in a sealed chamber [Holland, 1973; McGuire, 1990]. These elements can be gears, lobes, screws, vanes, etc. These pumps do not require valves in order to operate. Pumping rates were varied by means of the shaft rotational speed. Specifications for these pumps were described by Santoyo-Gutierrez [1997]. Two Procon series 1300 and 2500 pumps were used with capacities up to $100 \, \text{gal h}^{-1}$ $(0.378 \, \text{m}^3 \, \text{h}^{-1})$ and from 115 to $240 \, \text{gal h}^{-1}$ $(0.435 \text{ to } 0.907 \, \text{m}^3 \, \text{h}^{-1})$ respectively at 1725 rpm with a maximum discharge pressure of 13.8 bar [Procon, 1981].

In order to drive the positive displacement Procon pumps using electronic frequency controllers, three-phase 220 V electric motors were installed.

5.6 Hydrostatic and Vacuum Tests

Once the AHPAEPU had been assembled, pressure tests using water were carried out to detect leaks in all the connections of the system. These hydrostatic pressure tests were made in two stages at 5.1 bar. The first stage consisted of pressurising water into the refrigerant solution circuit. In the second stage, the water was injected into the effluent circuit. Several leaks appeared in the thermocouple fittings, in the flowmeter bypasses and in the welding of the vessels. These leaks were subsequently corrected. Additional vacuum tests were carried out on the AHPAEPU. The vacuum tests were conducted at 600 mm Hg (0.789 bar) and several leaks appeared. These leaks were corrected until there were no leaks.

5.6.1   Instrumentation

The instrumentation system of the experimental AHPAEPU consisted of thermocouples, pressure transducers, pressure gauges, electronic balances, and frequency controllers which were installed and calibrated. Most of these components were connected to a data acquisition unit. This data acquisition unit was connected to a personal computer to control the operation of the AHPAEPU. A full description of all the components of this instrumentation system is presented elsewhere [Santoyo-Gutierrez, 1997]. Additionally, a 750 W backup power system supplied by the TRIPPLITE Co., USA, was used to avoid the loss of programming routines and the continuous data collection system caused by any electrical failures during the experimental runs.

5.6.2   Instrumentation system description

Figure 5.2 shows a detailed line diagram for the instrumentation in both the auxiliary systems and the experimental absorption heat pump. Figures 5.9 and 5.10 show a rear and front view of the AHPAEPU before it had been covered with insulation. Figures 5.11–5.13 show a rear, side and front view of the AHPAEPU after it had been fully instrumented and insulated.

5.6.3   Temperature

The experimental AHPAEPU was instrumented with type E (nickel/chromel/ constantan) thermocouples. The temperature range is −200 to 900°C (−8.824 to 68.763 mV) with a maximum error of 1.7% for temperatures higher than 0°C [Omega, 1994]. Thirty thermocouples were prepared and welded. The location of these thermocouples is shown in Figure 5.2. These thermocouples were connected to the 3497 data acquisition/control unit supplied by the Hewlett/Packard Co., USA.

5.6.4   Pressure

In order to measure pressure, a barometer, vacuum gauges and differential pressure transducers were installed in the AHPAEPU. Figure 5.2 shows the physical distribution of the pressure sensors.

Atmospheric pressure was measured using a D469 mercury column barometer supplied by the Princo Nova™ Co., USA. The atmospheric pressure measurements were obtained with an accuracy of 0.1 mm Hg (0.013 kPa).

Figure 5.9 Physical distribution of the main components in the experimental absorption heat pump assisted effluent purification unit before it had been covered with insulation (rear view).

Figure 5.10 Physical distribution of the main components in the experimental absorption heat pump assisted effluent purification system before it had been covered with insulation (front view).

Five Bourdon gauges were installed as vacuum and pressure indicators in the AHPAEPU (Figure 5.2).

Differential pressure transducers

Five differential pressure transducers supplied by the Validyne Engineering Co., USA, were installed on the main vessels of the AHPAEPU. Four of the

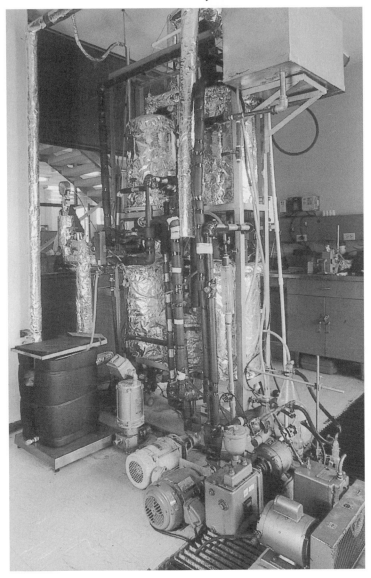

Figure 5.11 Physical distribution of the main components in the experimental absorption heat pump assisted effluent purification unit after it had been fully instrumented and insulated (rear view).

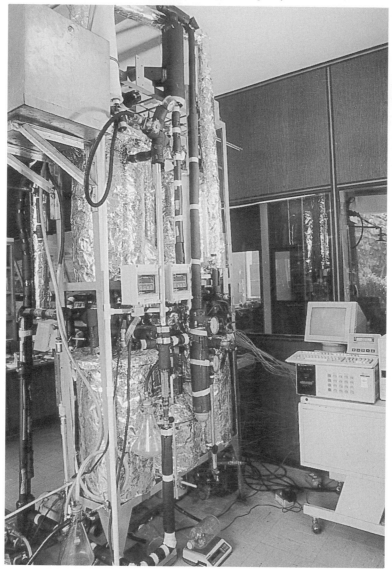

Figure 5.12 Physical distribution of the main components in the experimental absorption heat pump assisted effluent purification unit after it had been fully instrumented and insulated (side view).

Figure 5.13 Physical distribution of the main components in the experimental absorption heat pump assisted effluent purification unit after it had been fully instrumented and insulated (front view).

pressure transducers were chosen to measure vacuum conditions (type DP15/42). The remaining differential pressure transducer was placed on the separator steam supply line of the AHPAEPU (type DP15/52). These transducers were coupled to a MC170 rack supplied by the Validyne Engineering Co., which was connected to the 3497 data acquisition/control unit supplied by the Hewlett/Packard Co., USA, to record the pressure data during the experimental runs.

5.6.5   Data logging system

A screen display and the computer system to show and to monitor the operation, in real time, of the AHPAEPU were designed and installed. This system was based on a data acquisition unit and an IBM-compatible personal computer (PC). This system consisted of a 3497 data acquisition/control unit supplied by the Hewlett/Packard Co., USA, connected to a PC through an AT-GPIB interface board supplied by the National Instruments Co., USA. This interface enabled all the data monitored to be saved and stored in the PC's hard disk each thirty seconds.

5.6.6   Mass flow rate

In order to evaluate the flow rate in the different circuits of the AHPAEPU, two digital flowmeters and three frequency controllers were installed.

There were two refrigerant solution circuits in the AHPAEPU. In the first, the refrigerant-poor solution left the bottom of the generator and entered the top of the absorber. In the second, the refrigerant-rich solution left the bottom of the absorber and entered the top of the generator. An important parameter involved in the absorption process of the AHPAEPU is the flow ratio of the refrigerant solution. This is based on the flow rates in both the previously discussed circuits.

Flowmeters

Two digital flowmeters from Platon Flow Control Ltd, UK, were installed, one on each of those circuits. Specifically, these flowmeters were Platon gap meter type SDF series 2000 model 2044 [Platon, 1994].

Electronic frequency controllers

An electronic frequency controller is capable of reducing the frequency of the electric motor to as low as 5 Hz and therefore it can govern the shaft rotational speed of the electric motor through its magnetic fields. For this reason, the combinations consisting of (a) an electronic frequency controller, (b) a three-phase 220 V electric motor and (c) a positive displacement pump were installed in the absorption heat pump assisted effluent purification unit to substitute for flowmeters in each circuit of the AHPAEPU.

5.6.7 Weighing system

In order to calibrate the flow rate capacity of the Procon pumps, the fluid volume was weighed against time. Two electronic balances from the Ohaus Scale Co., USA, were used in the calibration tests with a maximum weight capacity up to 66 kg which allows it to record weights in the high speed range [Santoyo-Gutierrez, 1997].

Differential pressure transducer calibration procedure

The calibration of the differential pressure transducer installed on the steam separator supply line was made using a 35260/2 dead-weight balance supplied by the Dewit Co., USA, together with the voltmeter built into the 3497 data acquisition/control unit supplied by the Hewlett/Packard Co., USA.

5.6.8 Concentration of the salt solution

Water/lithium bromide solution

The lithium bromide solution was prepared using distilled water and the appropriate amount of salt. A calibration curve was constructed of salt concentration against the refractive index based on several samples containing from 45% to 60% by weight of lithium bromide. These samples were placed in an Abbe refractometer from Bellingham and Stanley Ltd, UK, at 40°C. The corresponding readings were converted to refractive indexes. These values were used to obtain, by polynomial regression, the equation that relates the concentration of salt solution from the refractive index as follows:

$$X = -1815.62\,ir^2 + 5557.91\,ir - 4187.31 \qquad (5.1)$$

where X = concentration of salt solution, (%wt) and ir = refractive index.

Water/Carrol solution

The water/Carrol solution was prepared using distilled water and the appropriate amount of lithium bromide and ethylene glycol. A calibration curve was constructed of the Carrol concentration against the refractive index based on several samples containing from 50% to 70% by weight of Carrol. These samples were placed in an Abbe refractometer from Bellingham and Stanley Ltd., UK, at 40°C. The corresponding readings were converted to refractive indexes. These values were used to obtain by polynomial regression, the equation that relates the concentration of salt solution from the refractive index as follows:

$$X = -784.432 \, ir^2 + 2569.39 \, ir - 2017.37 \qquad (5.2)$$

where X = concentration of Carrol solution (%wt) and ir = refractive index.

5.7 Experimental procedure

The system was loaded with 22 kg of salt solution at 52% of lithium bromide (LiBr) of 99.9% purity (dry basis). Corrosion in the system was prevented by using sodium dichromate salt ($Na_2Cr_2O_7$) as a corrosion inhibitor in the salt solution. Prior to the salt solution loading, the air was completely removed from the refrigerant circuit by applying a vacuum of 600 mm Hg (80 kPa). Afterwards, the salt solution was loaded through the valve located in the bottom of the generator.

5.7.1 System start-up procedure

In order to have a standard procedure for all the experimental runs, a start-up procedure for the operation of the experimental AHPAEPU was developed.

Once the steam conditions were established for the experimental test, the electric boiler was turned on. The key to achieving a fast start-up of the electric boiler was maintaining the correct electrolyte concentration in the boiler water [Kewanee, 1984]. The typical time employed to provide a steam pressure of 830 kPa was 25 minutes. During this time, the main valve that provides steam to the distribution line must be closed. At the same time, the secondary valve located in the top of the boiler must be slightly open. This valve was used to evacuate the non-condensable gases in the boiler. Once the boiler working pressure was achieved, the main valve was opened slightly in order to avoid the pressure falling. The working pressure in all the pipelines was achieved in 20 minutes.

During the start up of the boiler, the air was completely removed from the AHPAEPU by applying a vacuum of 600 mm Hg (80 kPa). Afterwards, a 30% excess volume of water was loaded into the condenser to ensure a full charge of water in the refrigerant circuit.

The auxiliary equipment such as the Abbe refractometer, constant temperature bath, the lamp for the refractometer, the 3497 data acquisition/control unit supplied by the Hewlett/Packard Co., USA, the PC, the electronic frequency controllers, the electronic balances and the auxiliary condenser cooling water flow were turned on. As was mentioned in section 5.6.5, the data logging system was programmed to collect and record temperature and pressure measurements every thirty seconds on the PC's hard disk. The system was set into the monitor only mode to display all data in the PC's monitor during the initial heating. Once steady conditions were achieved, the data logging system was turned to the recording mode.

The atmospheric pressure was taken from a mercury column barometer which was used to convert the readings from gauge pressure to absolute pressure. After this, the vacuum pumps were turned on in order to evacuate the air completely from all the circuits. Later, the working vacuum conditions were established for each circuit in the system.

The salt solution was poured through a ball valve mounted at the bottom of the generator. Simultaneously, the control valve placed between the generator and the economiser was completely opened to transfer the salt solution to the absorber. This operation was carried out in order to check for any dirt particles in the secondary circuit that could clog the flowmeters or the solution pump. Afterwards, the solution pump located below the absorber was operated to circulate the salt solution to the generator. This operation was carried out until the same level was achieved in both vessels: generator and absorber. Also, the effluent pump and the refrigerant pump were turned on to circulate the liquid in their respective circuits.

A small amount of distilled water was added to the refrigerant circuit in the condenser to obtain rapid heating of the refrigerant in the evaporator. Thus the refrigerant vapour was absorbed by the refrigerant-poor solution in the absorber in order to achieve steady-state conditions as soon as possible.

The steam supply valve was opened allowing steam circulation through the coil in the generator. The amount of steam used by the generator was measured by weighing the condensate on a precision electronic balance.

The main parameters measured and controlled during the experimental runs were: (a) the steam flow rate, (b) the flow rates of the refrigerant-poor solution and the refrigerant-rich solution, (c) the temperatures, (d) the vacuum conditions, (e) the concentration of the salt solution, (f) the flow rate of the effluent, (g) the levels of the refrigerant-poor solution and the refrigerant-rich solution in the generator and absorber respectively and (viii) the cooling water flow rates. Later the effluent was circulated through the condenser to gain heat from the refrigerant vapour arriving from the evaporator. Then the heated effluent was passed through the absorber where a partial change of phase was achieved by the absorption process releasing its heat of solution in the absorber.

The two-phase effluent was then conducted into the effluent separator where the phase separation was carried out. Finally, the effluent vapour arrived at the evaporator where it was condensed to provide the heat of evaporation for the refrigerant. Finally, the pure condensed effluent in the evaporator was weighed in a precision electronic balance at room temperature.

Once the experimental test was completed, the steam supply valve was closed. After that, a small volume of refrigerant from the condenser was transferred to the absorber to provide a refrigerant rich solution. A reduction of salt concentration in the system was achieved by means of recirculating the refrigerant-rich solution between the absorber and the generator. Later, the salt solution pump was turned off. Simultaneously, the generator control valve was fully opened to completely transfer the refrigerant-poor solution into the absorber. This procedure was used to avoid any possibility of crystallisation of the salt solution in the generator.

Later, all the vacuum pumps were turned off. Special attention was paid to avoid contamination of the oil in the vacuum pumps by liquid from the vessels when the corresponding vacuum lines were purged. Thus, the valves installed in the vacuum lines were opened to allow atmospheric pressure into the complete system.

5.8   Experimental Results

The experimental work was done using the water/lithium bromide and the water/Carrol mixtures. Carrol is a mixture of lithium bromide and ethylene glycol [$(CH_2OH)_2$] in the ratio 4.5:1 by weight. The evaluation of the performance of the experimental AHPAEPU operating with both water/lithium bromide and water/Carrol mixtures was based on the values of the theoretical enthalpic coefficient of performance $(COP)_{EH}$ and the associated parameters.

The coefficient of performance of an absorption heat pump used for heating can be written in terms of enthalpies to give the enthalpic coefficient of performance

$$(COP)_{EH} = \frac{H_1 + [(FR) - 1]H_4 - (FR)H_5 + H_6 - H_8}{H_6 + [(FR) - 1]H_4 - (FR)H_5} \tag{5.3}$$

A heat pump effectiveness HPE for an absorption heat pump used for heating can be defined as the ratio of the actual coefficient of performance $(COP)_A$ to the enthalpic coefficient of performance $(COP)_{EH}$ by the equation

$$HPE = \frac{(COP)_A}{(COP)_{EH}} \tag{5.4}$$

The flow ratio (FR) is defined as the mass flow rate of the refrigerant-rich solution M_{AB} being pumped from the absorber to the generator to the mass flow rate of pure refrigerant M_R in the primary circuit.

$$(FR) = \frac{M_{AB}}{M_R} \tag{5.5}$$

(FR) can also be written in terms of the % salt concentrations by weight in the generator X_{GE} and the absorber X_{AB} as

$$(FR) = \frac{X_{GE}}{X_{GE} - X_{AB}} \tag{5.6}$$

It was considered that steady-state conditions were reached when the parameter readings remained constant for a period of 50 minutes.

Although the water/lithium bromide system has been extensively used and reported, the only experimental results reported in the literature for the water/Carrol system are on a small glass laboratory scale heat transformer [Rivera, 1996]. No data have been reported for conventional heat driven absorption heat pumps. Therefore, this is the first time that a systematic comparison has been made between the water/lithium bromide and the water/Carrol systems operating in heat driven absorption heat pumps.

Figures 5.14, 5.15 and 5.16 are plots of the actual coefficient of performance $(COP)_A$ against the gross temperature lift $(T_{CO} - T_{EV})$, the flow ratio (FR) and the production rate of the purified effluent respectively showing the differences between the two systems. These plots show that the water/Carrol system has significantly higher $(COP)_A$ values than the commonly used water/lithium bromide system.

This improvement is due to the fact that the water/Carrol system has a greater negative deviation from Raoult's law and a corresponding lower salt

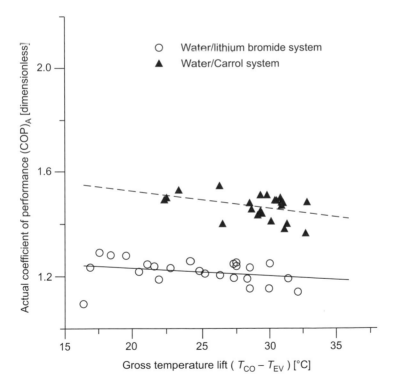

Figure 5.14 Plot of actual coefficient of performance $(COP)_A$ against the gross temperature lift $(T_{CO} - T_{EV})$ for the water/lithium bromide and the water/Carrol systems.

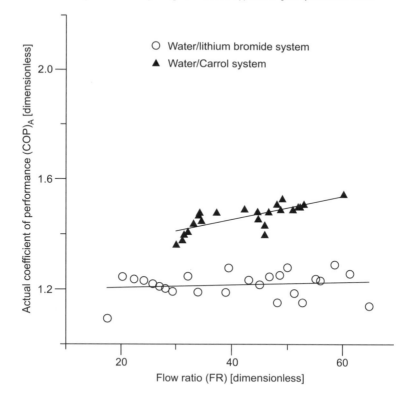

Figure 5.15 Plot of actual coefficient of performance (COP)$_A$ against the flow ratio (FR) for the water/lithium bromide and the water/Carrol systems.

concentration than the water/lithium bromide system. As a result, the water/Carrol system is able to give a higher condenser temperature T_{CO} for a given generator temperature T_{GE}.

This is probably because the ethylene glycol in the Carrol operates as a surfactant to enhance the heat and mass transfer processes in the liquid solution by reducing the surface tension forces. This in turn facilitates the absorption of water vapour refrigerant into the liquid absorbent solution. Additionally, mass transfer is enhanced at the vapour/liquid interface due to the increased turbulence. This interfacial turbulence is known as the Marangoni effect and is caused by surface tension gradients. These are ultimately related to the concentration and temperature dependence of the surface tension [Bjurström *et al.*, 1991; Herold, 1995].

Figures 5.17–5.19 are plots of the gross temperature lift ($T_{CO} - T_{EV}$), heat pump effectiveness (HPE)$_R$ and heat load Q respectively against the production rate of the purified effluent.

The most significant differences between the two systems are shown in Figures 5.18 and 5.19 respectively where the water/Carrol system has higher

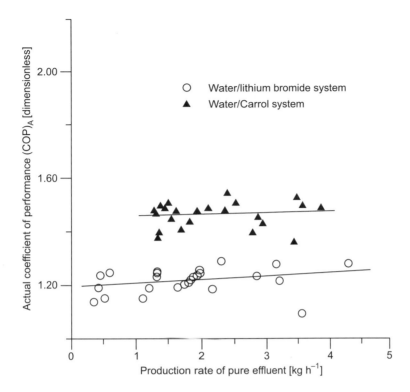

Figure 5.16 Plot of actual coefficient of performance $(COP)_A$ against the production rate of pure effluent for the water/lithium bromide and the water/Carrol systems.

values for the heat pump effectiveness and the heat load than for the conventional water/lithium bromide system. These results could provide the basis for the future development of more compact heat driven absorption heat pump systems with progressively lower capital costs.

5.9 Chemical Analyses

Chemical analyses were carried out to determine the chloride, sulphate and calcium concentrations. The effluent supplied, the effluent treated (distilled water) and the concentrated effluent were analysed. Table 5.1 shows the results of these analyses. The highest calcium concentration in the distilled water was $1.1 \, mg \, kg^{-1}$ but most of the results were less than this value. Similar behaviour was shown for the chloride and sulphate concentrations. The concentrated effluent had chloride concentrations ranging from 182.7 to $197.5 \, mg \, kg^{-1}$. The concentrations of other chemical species measured behaved in a similar manner increasing their concentration in the concentrated effluent.

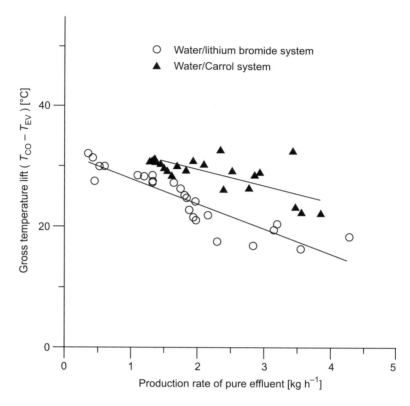

Figure 5.17 Plot of gross temperature lift ($T_{CO} - T_{EV}$) against the production rate of pure effluent for the water/lithium bromide and the water/Carrol systems.

The concentrations in the concentrated effluent could be used to estimate the feasibility of obtaining byproducts. The level of concentrations of such chemical substances was quite low in the distilled water. Similar quality distilled water was obtained in a laboratory still using tap water.

The pure water produced by the prototype AHPAEPU unit had a similar quality to the commercially available distilled water in Mexico with concentrations less than $1\,\text{mg}\,\text{kg}^{-1}$ for chlorides, calcium and sulphates.

5.10   Corrosion Testing Study

A comprehensive corrosion study has been carried out using aluminium, carbon steel, copper, stainless steel and cupro/nickel samples immersed in water/lithium bromide and water/Carrol solutions. The maximum corrosion rate produced by both fluids was $0.5\,\text{mm}\,\text{y}^{-1}$ on the aluminium samples. The corrosion was very limited in the case of the remaining materials. Additionally, there were no significant differences in corrosivity between the water/lithium bromide and the water/Carrol solutions.

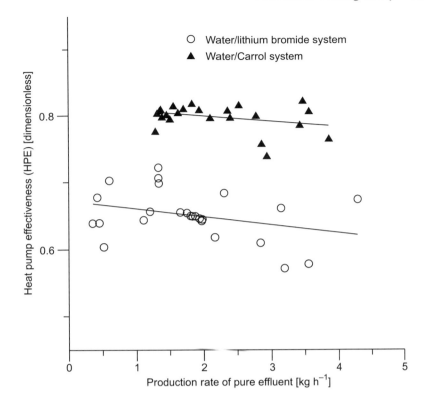

Figure 5.18 Plot of heat pump effectiveness (HPE) against the production rate of pure effluent for the water/lithium bromide and the water/Carrol solutions.

The study of the behaviour of materials subject to corrosive environments is an essential contribution to the correct choice of the materials to be used to construct the main components of industrial equipment. Corrosion studies are important in the context of: (a) cost reduction, (b) equipment and human safety, and (c) energy conservation. Costs can be reduced by decreasing or controlling the loss of materials through corrosion. The selection of materials which are corrosion resistant will reduce the risk of equipment failure. The latter can cause accidents which can also have an adverse effect on profitability. Some of these corrosion-related costs are:

- replacement of corroded equipment;
- overdesign to tolerate corrosion;
- plant shutdowns caused by corrosion failure;
- product contamination or loss;
- decreased process efficiency.

The prevention of the loss of materials, due to corrosion, will also save energy since resources including energy are employed to produce and fabricate

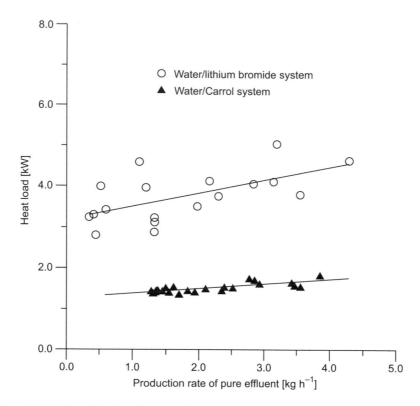

Figure 5.19 Plot of heat load against the production rate of pure effluent for the water/
lithium bromide and the water/Carrol systems.

materials. The corrosion of a metal is defined as a destructive attack by means
of a chemical reaction or electrochemical reaction with its environment [Van
Vlack, 1970]. The chemical nature of a material plays an important role in
the context of the surrounding environment and the product of the reaction
itself. For example, when copper and aluminium are corroded they produce
low permeability products [Seabright and Fabian, 1963]. These corrosion
products have an effect similar to that produced by paint protecting the metal
surface. When the corrosion products are formed, free energy is released as
illustrated by the following equation:

$$2Fe^0 + \tfrac{3}{2}O_2 \rightarrow Fe_2O_3 + \Delta G^0 = -741 \, kJ \, mol^{-1} \tag{5.7}$$

The negative free energy of formation indicates the tendency of the metal to
react, i.e. the oxide is stable. A positive free energy of formation indicates that
the metal in its base state is stable. In general, there are three main corrosion
forms. The first form is uniform corrosion, the second is localised corrosion,
and the third is galvanic corrosion.

Table 5.1 Results of the chemical analyses of the effluent supplied, the distilled water and the concentrated effluent from the experimental heat pump assisted purification unit concentration in $mg\,kg^{-1}$

Test number	Effluent supplied			Effluent treated			Effluent concentrated		
	Chlorides	Calcium	Sulph-ates	Chlorides	Calcium	Sulph-ates	Chlorides	Calcium	Sulph-ates
1	137.3	150.8	59.2	0.2	1.1	0.4	197.5	217.0	85.2
2	128.4	141.9	50.3	0.1	1.0	0.4	184.7	204.2	72.4
3	131.5	145.0	53.4	0.2	1.0	0.4	189.2	208.6	76.8
4	134.4	147.9	56.3	0.2	1.0	0.4	193.4	212.8	81.0
5	133.6	147.1	55.5	0.2	1.0	0.4	192.2	211.6	79.8
6	131.3	144.8	53.2	0.2	1.0	0.4	188.9	208.3	76.5
7	129.5	143.0	51.4	0.2	1.0	0.4	186.3	205.7	74.0
8	134.6	148.1	56.5	0.2	1.0	0.4	193.7	213.1	81.3
9	136.7	150.2	58.6	0.2	1.1	0.4	196.7	216.1	84.3
10	127.0	140.5	48.9	0.1	1.0	0.3	182.7	202.1	70.4
11	128.7	142.2	50.6	0.1	1.0	0.4	185.2	204.6	72.8
12	130.0	143.5	51.9	0.2	1.0	0.4	187.0	206.5	74.7
13	130.4	143.9	52.3	0.2	1.0	0.4	187.6	207.0	75.2
14	128.9	142.4	50.8	0.1	1.0	0.4	185.5	204.9	73.1
15	133.7	147.2	55.6	0.2	1.0	0.4	192.4	211.8	80.0

Uniform corrosion is the most common type of corrosion caused by acids such as carbonic or others. Uniform corrosion causes a metal to corrode at the same rate over its entire surface. This type of corrosion can be easily detected and measured. In this case, corrosion failures can be avoided by means of a continuous surface inspection.

In localised corrosion, a non-uniform corrosion type is presented as a local accelerated attack on small and specific areas. This attack is shown by a pit or cavity on the metal surface. A deeper penetration in this pit produces a rapid failure in this small area. The detection and control process for localised corrosion is difficult because there are several types of localised corrosion.

Galvanic corrosion is presented when different metals are exposed to a solution capable of carrying an electric current. The electric current is originated by a difference in potential associated between both metals.

5.10.1 Corrosion tests in water/lithium bromide and water/ Carrol solutions

In order to obtain the corrosive profiles for the main working fluids used in the AHPAEPU, an immersion corrosion testing programme was developed. The information generated from these corrosion tests will be very useful for scale up to industrial size units. Additionally, it will contribute to a comprehensive corrosion data bank, since there are no relevant quantitative corrosion data available in the open literature. Henthorne [1972] listed the essential reasons for carrying out a comprehensive corrosion testing programme as follows:

- to estimate the corrosion under simulated conditions as it occurs in a plant;
- to assess material and environmental effects for future applications;
- to review the quality of a specific material of known performance;
- to determine the mechanism of corrosion.

One of the disadvantages of absorption heat pumps is that the working fluids used tend to be corrosive. In order to minimise the cost, special attention has been paid to those materials that would be most appropriate for the construction of industrial-scale absorption heat pumps. The planning of a corrosion study was considered to be an integral part of the experimental study of the AHPAEPU constructed at the Instituto de Investigaciones Eléctricas (IIE), Cuernavaca, Mexico.

Carbon steel and stainless steel 316 were the materials using in the construction of the AHPAEPU [Santoyo-Gutierrez, 1997].

In order to obtain uniform corrosion rates, it was planned to test four specimens of the same construction material in each ageing cell. Therefore, the number of cells to be used in each test was five. It is important to comment that this number was doubled since it was planned to have two exposure times. The exposure times were one month and two months. The concentration of corrosive fluid, the agitation rate and temperature were maintained constant during each corrosion test.

5.10.2   Corrosion results

The highest average corrosion rates were observed for the aluminium specimens exposed to the water/lithium bromide, the water/lithium bromide with sodium dichromate and the water/Carrol solutions. The macroscopic and microscopic observations showed comprehensive damage over the entire surface of the aluminium specimens. Generalised deep pitting occurred on the surface which produced massive corrosion deposits. The corrosive fluids were severely contaminated due to the strong reaction by the specimens. Therefore, this material is not recommended for use in the presence of such corrosive fluids.

The lowest average corrosion rates were noted for the cupro/nickel and the stainless steel specimens. Contamination of the corrosive fluids was negligible which indicates a low reaction between the specimens and the corrosive fluids. The microscopic observations showed only colour changes on their metallic surfaces. Therefore, these materials are suitable for use in the presence of the corrosive fluids used in this corrosion study. However, in order to detect any type of corrosion, it is suggested that an additional corrosion control programme should be carried out.

High average corrosion rates were obtained for the carbon steel specimens exposed to the water/lithium bromide, the water/lithium bromide with sodium dichromate and the water/Carrol solutions. The microscopic observations showed a serious attack over the entire surface of the carbon steel specimens. Also, deep pitting on their surfaces was noted. The corrosive fluids were contaminated due to the moderate reaction by the specimens. This material is not widely recommended. However, it can be used with an appropriate thickness of material. Additionally, a systematic corrosion control programme should be followed.

The low average corrosion rates for the copper specimens exposed to the water/lithium bromide, the water/lithium bromide with sodium dichromate and the water/Carrol solutions were noted. The microscopic observations showed an initial attack on the surface of the copper specimens. No deep pitting on the their surfaces was noted. The corrosive fluids were contaminated but the reaction by the specimens was low. The average corrosion rates obtained for the copper specimens look very interesting, especially if availability and cost factors are considered. However, in order to detect and reduce the effect of any corrosion problems, it is suggested that a corrosion control programme should be always be carried out.

5.11 Estimated Cost of an Absorption Heat Pump Assisted Effluent Purification Unit

The cost of an absorption heat pump depends largely on the size and type of the heat exchangers. There are six heat exchangers in the AHPAEPU. These are the generator, absorber, condenser, evaporator, auxiliary condenser and economiser (solution heat exchanger). The first five exchange heat and mass between the cycle and its surroundings, while the economiser transfers heat internally within the cycle. Table 5.2 lists the main parameters and the heat transfer surface areas for each heat exchanger.

Once the surface area, tube length, shell diameter and number of tubes are known, the approximate cost of a shell and tube exchanger can be determined

Table 5.2 Heat transfer areas for the heat exchangers, in the experimental heat pump assisted purification unit

Component	ΔT_{lm} (°C)	U^* ($Wm^{-2}°C^{-1}$)	Q (kW)	A (m^2)
Generator	44.6	293	3.75	0.29
Absorber	11	258.5	2.87	1.01
Condenser	4.8	5200	0.39	0.02
Evaporator	2	1004	0.84	0.43
Auxiliary condenser	32	1004	1.33	0.04
Economiser	26.5	530	0.6	0.04

Sum of the total heat transfer areas = 1.83 m^2

---

* Values from Serpente *et al.* [1994].
Equipment cost US$75.5 per m^2 for 316 stainless steel. The values of Q (kW) depend upon the temperature differences ΔT_{lm} and so the values of ΔT_{lm} need to be representative of those for the pilot rig.

by using the procedures described in Perry *et al.* [1984] and by Peters and Timmerhaus [1991]. The simplest factorial method for estimating the fixed capital C_{FC} of a plant based on design is the Lang factor method [Holland *et al.*, 1983] using the following equation.

$$C_{FC} = f_L \sum C_{EQ} \tag{5.8}$$

where $f_L = 4.74$ for fluid processing and $\sum C_{EQ}$ is the sum of the delivered costs of all the major items of process equipment. In order to be on the conservative side, the cost for the most expensive construction material, i.e. 316 stainless steel, was used. The fixed capital cost (C_{FC}) obtained for the AHPAEPU was approximately US$30 000. This is consistent with current quotations obtained in Mexico.

5.12   References

Bjurström, H., Yao, W. and Setterwall, F. (1991) Heat transfer additives in absorption heat pumps, *Studsvik Energy, Report Studsvik/ED/90/33*.

Frias, J. L. (1991) An experimental study of a heat pump assisted purification system for geothermal brine, MSc Thesis, University of Salford, UK.

Frias, J. L., Siqueiros, J., Fernandez, H., Garcia A. and Holland, F. A. (1991) Developments in geothermal energy in Mexico - Part 36: The commissioning of a heat pump assisted brine purification system, *Heat Recovery Systems*, **11**(4), 297–310.

Henthorne, M. (1972) The first step in materials selection: Understanding corrosion, *Chemical Engineering/Deskbook Issue*, December 4, 19–30.

Herold, K. E. (1995) Design challenges in absorption chillers, *Journal of Mechanical Engineering*, **80**, October.

Holland, F. A. (1973) *Fluid Flow for Chemical Engineers: SI Units*, Edward Arnold, London, UK, pp. 58–77.

Holland, F. A., Watson, F. A. and Wilkinson, J. K. (1983) *Introduction to Process Economics*, 2nd edn, John Wiley, New York, USA.

Kestin, J., Dipippo, R., Khalifa, E. and Ryley, J. (1980) *Sourcebook on the Production of Electricity from Geothermal Energy*, US Department of Energy, Stanford, Connecticut, USA.

Kewanee Boiler (1984) EBCOR Electrode Steam Boiler, Industrial Steam, a Division of Kewanee Boiler Corporation, PO Box 24084, Oakland, California, 94263, USA.

McGuire, J. T. (1990) *Pumps for Chemical Processing*, Marcel Dekker, New York, USA, pp. 105–37.

Omega (1994) *OMEGA Temperature Measurement Handbook and Encyclopaedia*, USA, **28**, Z12–Z19.

Perry, R. H., Green, D. and Maloney, J. O. (1984) Heat transfer equipment, in *Perry's Chemical Engineers' Handbook*, 6th edn, McGraw-Hill, New York, USA, pp. 11–29.

Peters, M. S. and Timmerhaus, K. D. (1991) *Plant Design and Economics for Chemical Engineers*, 4th edn, McGraw-Hill, New York, USA, pp. 615–17.

Platon (1994) *Gapmeter Type SDF Series 2000 Digital Flowmeter Model 2044: Operation and Maintenance Manual OMM 1004*, Platon Flow Control Ltd, Hampshire RG22 4PS, UK.

Procon (1981) *Procon Rotary Vane Type Pumps, Catalogue 2003*, Procon Products Co., Murfreesboro, Tennessee, USA.

Rivera, W. (1996) Heat Transformer technology and steam generation, PhD Thesis, University of Salford, UK.

Santoyo-Gutierrez, S. (1997) Absorption heat pump assisted effluent purification, PhD Thesis, University of Salford, UK.

Seabright, L. H. and Fabian, R. J. (1963) The many faces of corrosion, *Materials Design Engineering*, January, 85–91.

Serpente, C. P., Kernen, M., Seewald, J. S. and Perez-Blanco, H. (1994) A 2 kW lithium bromide absorption machine with heat recovery and recirculation for novel fluid testing, in *Proceedings of the International Absorption Heat Pump Conference, New Orleans, USA*, January 19–21, AES-31, 65–71.

Siqueiros, J., Heard, C. and Holland, F. A. (1995) The commissioning of an integrated heat-pump assisted geothermal brine purification system, *Heat Recovery Systems*, 15(7), 655–64.

Siqueiros, J., Fernandez, H., Heard, C. and Barragan, D. (1992) Desarrollo e implantacion de tecnologia de bombas de calor, *Final Report: INFORME IIE/FE/11/2963/F*, Cuernavaca, Mexico.

Van Vlack, L. H. (1970) *Materials Science for Engineers*, Addison-Wesley, Reading, Massachusetts, USA.

6 Economics of heat pump systems

6.1 Nomenclature

A	annual cost or payments [US\$ y^{-1}]
A_B	annual operating cost of a gas boiler [US\$ y^{-1}]
A_{BD}	annual balance sheet depreciation charge [US\$ y^{-1}]
A_{CI}	annual cash income [US\$ y^{-1}]
A_{CF}	net annual cash flow [US\$ y^{-1}]
A_D	annual amount of depreciation [US\$ y^{-1}]
A_{DCF}	net annual discounted cash flow [US\$ y^{-1}]
A_{DME}	annual direct manufacturing cost or expense [US\$ y^{-1}]
A_{FC}	annual fixed capital cost of a compressor driven heat pump [US\$ $y^{-1}kW^{-1}$]
A_{GE}	annual general expense [US\$ y^{-1}]
A_{GP}	annual gross profit [US\$ y^{-1}]
A_{HP}	annual operating cost of a gas driven heat pump [US\$ y^{-1}]
A_{IT}	annual amount of tax [US\$ y^{-1}]
A_{IME}	annual indirect manufacturing cost or expense [US\$ y^{-1}]
A_{ME}	annual manufacturing cost or expense [US\$ y^{-1}]
A_{NCI}	net annual cash income [US\$ y^{-1}]
A_{NP}	net annual profit [US\$ y^{-1}]
A_{NNP}	net annual profit after tax [US\$ y^{-1}]
A_S	revenue from annual sales [US\$ y^{-1}]
A_{TC}	annual expenditure of capital [US\$ y^{-1}]
A_{TE}	total annual cost or expense [US\$ y^{-1}]
A_{TFE}	total annual fixed expense [US\$ y^{-1}]
A_{TVE}	total annual variable expense [US\$ y^{-1}]
c_B	unit cost of base heating [US\$ kWh^{-1}]
c_D	unit cost of heat energy delivered [US\$ kWh^{-1}]
c_E	unit cost of electricity [US\$ kWh^{-1}]
c_i	unit cost of input energy to a compressor [US\$ kWh^{-1}]
c_L	labour cost [US\$$h^{-1}$]
c_{LC}	labour element of unit cost of water production [US cents $kg^{-1} h^{-1}$]

c_S	sales price [US\$ unit^{-1}]
c_{st}	unit cost of steam heating [US\$ kWh^{-1}]
c_{TVE}	total variable expense per unit [US\$ unit^{-1}]
c_W	capital element of unit cost of water purification [US\$ kg^{-1}]
C_B	fixed capital cost of a gas boiler [US\$]
C_{EQ}	equipment cost [US\$]
C_{FC}	fixed capital cost [US\$]
$\bar{C}_{FC}$	normalised fixed capital cost of a compressor driven heat pump [US\$ kW^{-1}]
C_{HP}	fixed capital cost of a gas driven heat pump [US\$]
C_{TC}	total capital cost [US\$]
C_{WC}	working capital cost [US\$]
C_L	cost of land and other nondepreciables [US\$]
$(COP)_A$	actual coefficient of performance [dimensionless]
$(DCFRR)$	discounted cash flow rate of return [fraction or % y^{-1}]
f_{AF}	annuity future worth factor [dimensionless]
f_{AP}	annuity present worth factor [dimensionless]
f_d	discount factor [dimensionless]
f_i	compound interest factor [dimensionless]
f_L	Lang factor [dimensionless]
F	future worth of a sum of money [US\$]
i	fractional annual interest rate on borrowed money [y^{-1}]
M	production rate [kg h^{-1}]
n	number of years [dimensionless]
(NPV)	net present value [US\$]
P	present worth of a sum of money [US\$]
(PER)	primary energy ratio [dimensionless]
(PBP)	payback period [yr]
Q_D	heat delivered by a heat pump [kW$_t$]
r	ratio of unit cost of heat to work input [dimensionless]
R	annual production rate [units y^{-1}]
R_B	annual break-even production rate [units y^{-1}]
S	scrap value [US\$]
W	rate of work input [kW]
y	number of operating hours per year [y^{-1}]
γ	ratio of unit cost of work input to heat [dimensionless]

6.2   Introduction

Although heat pumps have a huge potential to save energy, they have up to the present been judged on their potential to save money.

Heat pump systems for refrigeration have developed much more rapidly than heat pump systems for heating. This is because there is no alternative technology for refrigeration, whilst heat pump systems for heating have had to compete against a variety of other well-developed heating systems.

Furthermore, early heat pump systems had some reliability problems and often were not well matched to their applications. With the increasing availability of more accurate design data and more sophisticated equipment and controls, these problems are now a thing of the past. However, there still remains a considerable potential to benefit from the lessons learnt in order to produce heat pump systems with a lower capital cost. This is particularly true in the case of the heat driven absorption systems.

The economics of mechanical vapour compression heat pumps are critically dependent on the unit cost of the input energy to the compressor and the ratio of this to the unit cost of alternative heat energies, such as steam. In contrast, the economics of heat driven absorption systems depend very largely on the capital cost, which can be expected to be reduced significantly in the future as a result of the benefits of the lessons learnt.

In several countries with high energy costs, important industrial developments are underway. A number of large-scale heat driven absorption heat transformer systems, which produce steam for industrial processes, are currently operating in Japan with payback periods of under two years. Similar developments are taking place in Sweden.

Heat driven absorption systems have the further advantage of operating with environmentally clean working fluids.

In the case of mechanical vapour compression heat pumps, a comprehensive range of environmentally friendly working fluids is still being developed to replace traditional chlorofluorocarbon working fluids. Additionally, if the electricity used to drive the compressors is obtained from the combustion of fossil fuels, then this involves an associated pollution.

Stricter government regulations with severe economic penalties for polluting the environment could greatly increase the economic advantages of using heat driven absorption heat pump systems in the future.

6.3   Elements of Profitability Assessment

6.3.1   Time value of money

Interest payable on borrowed money is the basis of modern business activity. When money is loaned, there is always a risk that it may not be returned. Interest is the inducement offered to make the risk acceptable. The amount of a loan P is known as the principal. The future worth F of the money is greater than its present worth P. The relationship between F and P depends on the type of interest used.

When simple interest is used F and P are related by the equation

$$F = P(1 + ni) \tag{6.1}$$

where $i =$ fractional interest rate per period, and $n =$ number of interest periods.

The interest period is commonly chosen to be one calendar year. More commonly, compound interest is used where F and P are related by the equation

$$F = P(1+i)^n \qquad\qquad (6.2)$$

or

$$F = Pf_i \qquad\qquad (6.3)$$

where $f_i = (1+i)^n$ is the compound interest factor, values of which are readily available in tables similar to those in Appendix 5.

The present value P of a future sum of money F is given by the equation

$$P = \frac{F}{(1+i)^n} \qquad\qquad (6.4)$$

or

$$P = Ff_d \qquad\qquad (6.5)$$

where

$$f_d = \frac{1}{f_i} = \frac{1}{(1+i)^n}$$

is the discount factor, values of which are readily available in tables similar to those in Appendix 5.

The proof of the annual compound interest formula is as follows. After the first year the worth is

$$F(1) = P + Pi = P(1+i)$$

After the second year the worth is

$$\begin{aligned} F(2) &= P(1+i) + P(1+i)i \\ &= P(1+i)^2 \end{aligned}$$

After the third year the worth is

$$\begin{aligned} F(3) &= P(1+i)^2 + P(1+i)^2 i \\ &= P(1+i)^3 \end{aligned}$$

After the nth year the worth is

$$F(n) = P(1+i)^n$$

which is normally written as

$$F = P(1 + i)^n \tag{6.2}$$

A series of equal annual payments A invested at a fractional rate i and made at the end of each year over a period of n years may be used to build up a future sum of money F given by the equation

$$F = A \left[\frac{(1 + i)^n - 1}{i} \right] \tag{6.6}$$

or

$$F = \frac{A}{f_{AF}} \tag{6.7}$$

where

$$f_{AF} = \frac{i}{(1 + i)^n - 1}$$

is the annuity future worth factor, values of which are readily available in tables similar to those in Appendix 5.

Equation (6.6) can be combined with equation (6.4) to give

$$P = A \left[\frac{(1 + i)^n - 1}{i(1 + i)^n} \right] \tag{6.8}$$

or

$$P = \frac{A}{f_{AP}} \tag{6.9}$$

where P is the present worth of the series of future equal annual payments A and

$$f_{AP} = \frac{i(1 + i)^n}{(1 + i)^n - 1}$$

is the annuity present worth factor, values of which are readily available in tables similar to those in Appendix 5.

Figure 6.1 is a plot of the annuity present worth factor f_{AP} against the annual compound interest rate for various years of life.

Alternatively the annual payment A required to build up a future sum of money F with a present value of P is given by the equations

$$A = F f_{AF} \tag{6.10}$$

or

$$A = P f_{AP} \tag{6.11}$$

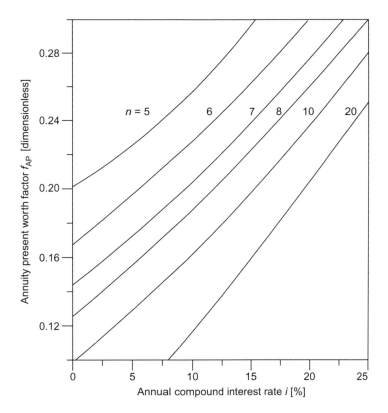

Figure 6.1 Plot of annuity present worth factor against the annual compound interest rate for various years of life.

The proof of these equations is as follows.

The future worth of the first annual payment A is

$$F = A(1 + i)^{n-1} \tag{6.2}$$

after compounding for $n - 1$ years.

The future worth of the second annual payment A is

$$F = A(1 + i)^{n-2} \tag{6.2}$$

after compounding for $n - 2$ years.

Thus the combined future worth of all the payments is

$$F = A\left[(1 + i)^{n-1} + (1 + i)^{n-2} + (1 + i)^{n-3} + \cdots + (1 + i)^{n-n}\right]$$

where $(1 + i)^{n-n} = 1.$

Multiply both sides of the above equation by $(1 + i)$ to give

$$F(1 + i) = A\left[(1 + i)^n + (1 + i)^{n-1} + (1 + i)^{n-2} + \cdots + (1 + i)\right]$$

Subtract the previous equation from this to give

$$F(1 + i) - F = A[(1 + i)^n - 1]$$

which can be rewritten as

$$F = A\left[\frac{(1 + i)^n - 1}{i}\right] \tag{6.6}$$

6.3.2   Annual cash flows and costs

The revenue A_S from the annual sales of a product minus the total annual cost or expense A_{TE} required to produce and sell the product excluding any annual provision for plant depreciation is the annual cash income A_{CI}.

$$A_{CI} = A_S - A_{TE} \tag{6.12}$$

The net annual cash income A_{NCI} is the annual cash income minus the annual amount of tax A_{IT}.

$$A_{NCI} = A_{CI} - A_{IT} \tag{6.13}$$

The net annual cash flow after tax A_{CF} is given by the equation

$$A_{CF} = A_{NCI} - A_{TC} \tag{6.14}$$

where A_{TC} is the annual expenditure of capital, which is not necessarily zero after the plant has been built. For example, plant additions or modifications may be required in future years. The net annual cash flow A_{CF} is the basis of the more modern methods of profitability assessment such as the net present value (NPV) method and the discounted cash flow rate of return (DCFRR) method. Here A_{CF} excludes any provision for depreciation. In both the (NPV) and (DCFRR) methods for profitability assessment, depreciation is inherently taken care of in the calculations which include capital recovery.

The total annual expense A_{TE} in equation (6.12) required to produce and sell a product can be written as the sum of the annual general expense A_{GE} and the annual manufacturing cost or expense A_{ME}.

$$A_{TE} = A_{GE} + A_{ME} \tag{6.15}$$

The annual general expense A_{GE} arises from the following items: administration, sales, shipping of product, advertising and marketing, technical service, research and development, and finance.

The terms gross annual profit A_{GP} and net annual profit A_{NP} are commonly used by accountants and businessmen. However they tend to be avoided by engineers because of their dimensional inconsistency.

The gross annual profit A_{GP} is given the equation

$$A_{GP} = A_S - A_{ME} - A_{BD} \qquad (6.16)$$

where A_{BD} is the annual balance sheet depreciation charge.

The net annual profit A_{NP} is simply A_{GP} minus the annual general expense A_{GE}.

$$A_{NP} = A_{GP} - A_{GE} \qquad (6.17)$$

Equation (6.16) can also be written as

$$A_{NP} = A_{CI} - A_{BD} \qquad (6.18)$$

The net annual profit after tax A_{NNP} can be written as

$$A_{NNP} = A_{NCI} - A_{BD} \qquad (6.19)$$

It should be noted that A_{CI} and A_{NCI} in equations (6.18) and (6.19) respectively are computed from actual cash flows whilst A_{BD} is merely a book transaction. This dimensional inconsistency is the reason for the unpopularity of the terms gross and net annual profits with engineers.

The annual manufacturing cost or expense A_{ME} can be written as the sum of the direct manufacturing cost or expense A_{DME} and the indirect manufacturing cost or expense A_{IME}.

$$A_{ME} = A_{DME} + A_{IME} \qquad (6.20)$$

The annual direct manufacturing cost or expense A_{DME} includes those expenses incurred directly from the production operation. These expenses arise from the following items: raw materials (including delivery), catalysts and solvents, operating labour, operating supervision, utilities, operating maintenance and repairs, operating supplies, royalties and patents. The direct manufacturing cost is also known as the prime cost.

The annual indirect manufacturing cost or expense includes those expenses incurred as a result of, but not directly from, the production operation. These expenses arise from the following items: payroll overhead, control laboratory, general plant overhead, packaging and storage facilities.

Other indirect costs include property taxes, rent and insurance. The indirect manufacturing cost A_{IME} is often referred to as the manufacturing overhead cost.

Accountants like to separate all costs into fixed and variable elements. The annual cash income A_{CI} given by equation (6.12) can also be written as

$$A_{CI} = A_S - A_{TVE} - A_{TFE} \tag{6.21}$$

where A_{TFE} and A_{TVE} are the total annual fixed expense and the total annual variable expense respectively.

If the annual sales volume can be taken as equal to the annual production rate R, the revenue A_S from annual sales is the product of R and the sales price c_S per unit of production.

$$A_S = Rc_S \tag{6.22}$$

Since the annual sales volume is strictly speaking R plus the inventory at the beginning of the year minus the inventory at the end of the year, equation (6.22) may be only approximately true. For the most part, the total annual variable expense A_{TVE} can be taken as proportional to the annual production rate R and the total variable expense c_{TVE} per unit of production.

$$A_{TVE} = Rc_{TVE} \tag{6.23}$$

Substitute equations (6.22) and (6.23) into equation (6.21) to give

$$A_{CI} = R(c_S - c_{TVE}) - A_{TFE} \tag{6.24}$$

where equation (6.24) gives the annual cash income A_{CI} in terms of the annual production rate R. The term $(c_S - c_{TVE})$ is called the contribution to cash income per unit of production where c_S is the sales price per unit.

The annual cash income A_{CI} is zero at an annual production rate R_B given by the equation

$$R_B = \frac{A_{TFE}}{(c_S - c_{TVE})} \tag{6.25}$$

The annual cash income can also be written by combining equations (6.24) and (6.25) as

$$A_{CI} = (R - R_B)(c_S - c_{TVE}) \tag{6.26}$$

6.3.3   Capital costs

The total capital cost C_{TC} of a plant consists of the fixed capital cost C_{FC} plus the working capital cost C_{WC} plus the cost of land and other non-depreciable costs C_L [Holland *et al.*, 1983].

$$C_{TC} = C_{FC} + C_{WC} + C_L \tag{6.27}$$

The project may be a complete plant or an addition to a plant such as a heat pump in an evaporation or distillation plant. It could also be a modification to an existing plant.

The fixed capital cost C_{FC} includes the following items:

- major process equipment (i.e. tanks, heat exchangers, pumps, compressors, etc.);
- installation of major process equipment;
- process piping;
- insulation;
- instrumentation;
- auxiliary facilities (i.e. power substations, transformers, boiler house, fire control equipment, etc.);
- outside lines (i.e. piping external to buildings plus supports and posts for overhead piping, electric feeders from power stations);
- land and site improvements;
- buildings and structures;
- engineering and construction (design and engineering fees plus supervision of plant erection);
- contractor's fees.

The working capital cost C_{WC} of a process includes the following items:

- raw materials for start-up;
- raw material, work in progress, and finished product inventories;
- cost of handling and transportation of materials to and from store;
- cost of inventory control, warehouse, associated insurance, security arrangements, etc.;
- money to carry accounts receivable (i.e. credit extended to customers), less accounts payable (i.e. credit extended by suppliers);
- money to meet payrolls when starting up;
- readily available cash for emergencies;
- any additional cash required to operate the process.

Since working capital is completely recoverable at any time, in theory if not in practice, no tax allowance is made for depreciation. The same is also true of the value of the land occupied by the plant. In the process industries, working capital is likely to be of the order of 10–20 % of the value of the fixed capital investment.

There are various types of capital cost estimates. These range from order of magnitude estimates which can be completed in a short time, to detailed estimates which require the collection of a considerable amount of data and the expenditure of much time and effort.

The fixed capital cost C_{FC} is usually defined as the capital required to provide all the depreciable facilities. It may be divided into two classes known as the battery limits and auxiliary facilities. The boundary for battery limits includes all manufacturing and processing equipment, whereas the auxiliary facilities include administrative offices, storage areas, utilities and other essential and non-essential supporting facilities.

Life-cycle costs take into account the original capital cost, the total operating and maintenance costs during service and the disposal costs and salvage value at the end of the operating life of the unit.

The main steps in preparing a capital cost estimation based on design are as follows:

1. initial idea for the process and specifications of the size and type of the operation;
2. collection of physical and chemical data from the literature, by prediction methods or from laboratory experiments;
3. preparation of a preliminary equipment flow sheet incorporating the required unit operations and showing the main items of equipment;
4. preparation of heat and mass balances;
5. specification of temperatures and pressures at various points on the equipment flow sheet;
6. design calculations to size the main items of equipment, such as pumps, compressors, heat exchangers, tanks, etc.;
7. preparation of a coded list of items of major process equipment, such as storage tanks, circulating and feed pumps, heat exchangers, compressors, etc., together with details giving size, capacity, materials of construction, operating pressures and temperatures, etc.;
8. collection of cost data from the literature, company records, or quotations;
9. estimation of the total delivered cost of all the items of major process equipment $\sum C_{EQ}$ from the coded list.

The simplest factorial method for estimating the fixed capital cost C_{FC} of a plant based on design is the Lang factor method [Lang, 1947, 1948] given by the equation

$$C_{FC} = f_L \sum C_{EQ} \tag{6.28}$$

where $f_L = 3.10$ for solids processing, $f_L = 3.63$ for mixed solids–fluid processing and $f_L = 4.74$ for fluids processing and $\sum C_{EQ}$ is the sum of the delivered costs of all the major items of process equipment. The omission of any necessary item of equipment will cause the estimate to be too low and the use of a check list is essential. The method is most effective in companies which build up their own Lang factors based on their own company data.

The fixed capital cost C_{FCI} for a plant with a production rate M_1 can be scaled up to give an estimate of the fixed capital cost C_{FC2} for a plant with a higher production rate M_2 using the seven-tenths power law given by the equation

$$\frac{C_{FC2}}{C_{FC1}} = \left(\frac{M_2}{M_1}\right)^{0.7} \tag{6.29}$$

Data from cost records of equipment purchased at an earlier date should be converted to values at a desired date by means of an appropriate cost index.

The simplest and most commonly used method of depreciation is the straight line method in which the average annual amount of depreciation A_D is given by the equation

$$A_D = \frac{(C_{FC} - S)}{n} \qquad (6.30)$$

where S is the scrap value of the plant after a life of n years.

6.3.4   Traditional methods for estimating profitability

The traditional rate of return methods for estimating profitability are based on the equation

$$\% \text{ rate of return} = \frac{\text{Annual return}}{\text{Invested capital}} \times 100 \qquad (6.31)$$

These methods are still widely employed by accountants in spite of the development of more modern methods for estimating profitability. Although the traditional methods based on equation (6.31) have the advantage of simplicity, they can give very misleading results.

The annual return in equation (6.31) can either be the annual cash income A_{CI} given by equation (6.12), the net annual cash income after tax A_{NCI} given by equation (6.13), the net annual profit A_{NP} given by equations (6.17) and (6.18) or the annual profit after tax A_{NNP} given by equation (6.19).

The invested capital in equation (6.31) can either be the original total capital cost, the depreciated capital value, the average capital value over the life of the plant, or the current replacement value. The depreciated capital value depends on the arbitrarily chosen method of depreciation. The total capital cost C_{TC} given by equation (6.23) includes the working capital C_{WC} and this in turn depends on the valuation of the inventory.

Since annual return and invested capital have so many meanings, it is necessary to define the terms precisely before making use of equation (6.31).

Another traditional method of measuring profitability is the payback period (PBP). However, this is not really a measure of profitability but of the time it takes for the cash flows to recoup the original fixed capital expenditure C_{FC}. The net annual cash flow after tax is given by the equation

$$A_{CF} = A_{NCI} - A_{TC} \qquad (6.14)$$

where A_{TC} is the annual expenditure of capital in a particular year, which is not necessarily zero after the plant has been built.

The payback period (PBP) is the time taken for the cumulative net cash flow $\sum A_{CF}$ from the start up of the plant to equal the depreciable capital investment C_{FC} where S is the scrap value of the plant. In the case of nuclear power plants, S has a negative value.

If the annual cash flows A_{CF} over a number of years can be assumed to be constant, as is the case with a typical energy saving project, the payback period (PBP) in years can be calculated by the following equation

$$(PBP) = \frac{C_{FC} - S}{A_{CF}} \tag{6.32}$$

The payback period (PBP) method takes no account of cash flows or profits received after the break-even point has been reached. The method is based on the premise that the earlier the fixed capital is recovered the better the project. However, this approach can be misleading.

Table 6.1 Comparison of the two projects on the basis of payback periods (US$)

Year	A_{CF} for project A	A_{CF} for project B
0	−100 000	−100 000
1	50 000	0
2	30 000	10 000
3	20 000	20 000
4	10 000	30 000
5	0	40 000
6	0	50 000
7	0	60 000
$\sum A_{CF}$	10 000	110 000
PBP	3 years	5 years

Consider project A and B with net annual cash flows listed in Table 6.1. Both projects have initial fixed capital expenditures of US$100 000. On the basis of payback period, project A is the most desirable since the fixed capital expenditure is recovered in three years compared with five years for project B. However project B runs for seven years with a cumulative net cash flow of US$110 000. This is obviously more profitable than project A which runs for only four years with a cumulative net cash flow of only US$10 000.

6.3.5   Profitability methods based on the time value of money

The present value P of a future sum of money F is given by equation (6.5)

$$P = Ff_d \tag{6.5}$$

where $f_d = 1/(1+i)^n$ is the discount factor. Thus, cash flow in the early years of a project has a greater value than the same amount of cash flow in the later years of the project. Time is taken into account by using the net annual discounted cash flow A_{DCF} which is related to the net annual cash flow A_{CF} and the discount factor f_d by the equation

$$A_{DCF} = A_{CF}f_d \qquad (6.33)$$

Thus, at the end of year 1

$$A_{DCF1} = \frac{A_{CF1}}{(1+i)}$$

At the end of year 2

$$A_{DCF2} = \frac{A_{CF2}}{(1+i)^2}$$

and at the end of year n

$$A_{DCFn} = \frac{A_{CFn}}{(1+i)^n}$$

The sum of the annual discounted cash flows over n years $\sum A_{DCF}$ is known as the net present value (NPV) of the project

$$(NPV) = \sum A_{DCF} \qquad (6.34)$$

The value of (NPV) is of course directly dependent on the choice of the fractional interest rate i. An interest rate can be chosen to make the net present value (NPV) = 0 after a chosen number of years. This value of i is given by the equation

$$(NPV) = \sum A_{DCF} = 0 \qquad (6.35)$$

and may be found either graphically or by an iterative trial and error procedure. The value i given by equation (6.35) is known as the discounted cash flow rate of return (DCFRR).

The main advantage of (DCFRR) over (NPV) is that it is independent of the zero or base year chosen. In contrast, (NPV) varies according to the zero year chosen. When the net present value (NPV) and discounted cash flow rate of return (DCFRR) are computed, depreciation is not considered as a separate factor. A (DCFRR) of, say, 15% implies that 15% per year will be earned on the investment in addition to which the project generates sufficient money to the repay the original investment. Frequently, depreciation is considered as the means of recovering the original investment. This is done implicitly in the case of (NPV) and (DCFRR) calculations.

For the simplified case of a single lump sum investment and equal annual cash flow savings for a system of infinite life, the payback period (PBP) is related to the fractional discounted cash flow rate of return (DCFRR) by the equation

$$(DCFRR) = \frac{1}{(PBP)} \tag{6.36}$$

Equation (6.36) shows that the maximum (DCFRR) which it is possible to reach for a project with a payback period of two years is 0.5 or 50%.

Equation (6.36) can be used to make rapid approximate estimates of (DCFRR) for long-term projects, such as energy savings projects.

6.4   Economics of Mechanical Vapour Compression Heat Pump Systems

Let the normalised fixed capital cost of a mechanical vapour compression heat pump system be $\overline{C}_{FC}$ in US\$ per kW of high-grade energy input. When interest charges are involved, $\overline{C}_{FC}$ can be related to an annual cost A_{FC} in US\$ per kW of high-grade energy input by the equation

$$A_{FC} = \overline{C}_{FC} f_{AP} \tag{6.37}$$

where

$$f_{AP} = \frac{i(1+i)^n}{(1+i)^n - 1}$$

is the annuity present worth factor, values of which are given in Appendix 5.

The plot of the annuity present worth factor against the annual compound interest rate for various years of life in Figure 6.1 indicates that f_{AP} is likely to lie in the range from 0.1 to 0.3 for most heat pump systems, with perhaps 0.2 as a typical value.

The high-grade energy input to the compressor of a mechanical vapour compression heat pump can be any form of mechanical energy, but it is most commonly electricity.

The unit cost of heat energy delivered by the heat pump c_D in US\$/kW h of heat output neglecting any maintenance cost, is given by the equation

$$c_D = \frac{c_i y + A_{FC}}{(COP)_A y} \tag{6.38}$$

where y is the number of operating hours per year, c_i is the unit cost of high-grade input energy to the compressor in US\$/kWh, and $(COP)_A$ is the actual coefficient of performance of the heat pump.

The actual coefficient of performance $(COP)_A$ of a compressor driven heat pump is given by the equation

$$(COP)_A = \frac{Q_D}{W} \tag{6.39}$$

where Q_D is the heat delivered by the heat pump in kWh_t per hour and W is the rate of mechanical work supplied by the compressor in kW.

The saving and heating cost by using a compressor driven heat pump in a particular process is $(c_B - c_D)$ in US\$/kWh where c_B is the unit cost of the base heating requirement when direct heating is used.

The payback period for a compressor driven heat pump is the additional fixed capital cost of the heat pump divided by the annual saving on heating costs. In terms of the normalised fixed capital cost $\overline{C}_{FC}$ in US\$/kW, the payback period (PBP) can be written either as

$$(PBP) = \frac{\overline{C}_{FC}}{(COP)_A y (c_B - c_D)} \tag{6.40}$$

or when combined with equation (6.38) as

$$(PBP) = \frac{\overline{C}_{FC}}{y[(COP)_A c_B - c_i] - A_{FC}} \tag{6.41}$$

The ratio of the unit cost of the base heat supply to the unit cost of the input energy to the compressor is

$$r = \frac{c_B}{c_i} \tag{6.42}$$

Since the value of r is normally significantly less than unity, it is sometimes more convenient to write it inversely as γ, the ratio of the unit cost of the input energy to the compressor to the unit cost of the base heat supply written as

$$\gamma = \frac{c_i}{c_B} \tag{6.43}$$

Equation (6.41) can be written in terms of equation (6.42) as

$$(PBP) = \frac{\overline{C}_{FC}}{c_i y [r(COP)_A - 1] - A_{FC}} \tag{6.44}$$

or in terms of equation (6.43) as

$$(PBP) = \frac{\overline{C}_{FC}}{c_i y \left[\frac{(COP)_A}{\gamma} - 1 \right] - A_{FC}} \tag{6.45}$$

For the special case of the unit cost of the base heat supply being the same as the unit cost of the input energy to the compressor, equations (6.44) and (6.45) can be written as

$$(PBP) = \frac{\overline{C}_{FC}}{c_i y\left[(COP)_A - 1\right] - A_{FC}} \tag{6.46}$$

Since all costs refer to a given year, equations (6.44)–(6.46) are independent of inflation. These equations show that in order to have a low payback period (PBP), the value of γ, $\overline{C}_{FC}$ and A_{FC} should be small and the values of $r(COP)_A$, c_i and y should be large.

Figure 6.2 is a qualitative plot of payback period against on-stream time for a mechanical vapour compression heat pump with a high and low coefficient of performance. The number of operating hours per year y is likely to be much larger for industrial process heat pump systems than for heat pumps used to

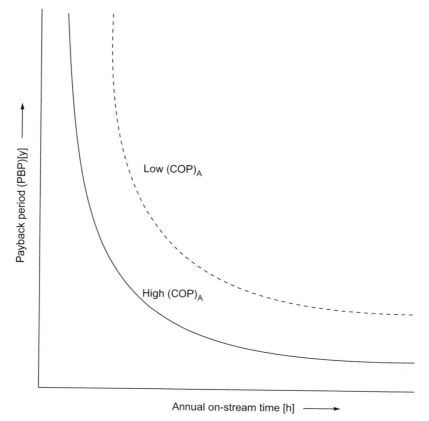

Figure 6.2 Plot of payback period against annual on stream time for a mechanical vapour compression heat pump with a high and low coefficient of performance.

heat buildings. Thus, industrial heat pump systems are likely to have significantly lower payback periods than those designed to heat buildings.

For a heat pump with a long life, saving equal amounts of energy each year, the fractional discounted cash flow rate of return (DCFRR) can be written as the inverse of the payback period (PBP) as in equation (6.36)

$$(\text{DCFRR}) = \frac{1}{(\text{PBP})} \tag{6.36}$$

Equations (6.44)–(6.46) can be rewritten in terms of equations (6.36) and (6.37) in their respective terms as

$$(\text{DCFRR}) = \frac{c_i y}{\overline{C}_{FC}} [r(\text{COP})_A - 1] - f_{AP} \tag{6.47}$$

$$(\text{DCFRR}) = \frac{c_i y}{\overline{C}_{FC}} \left[\frac{(\text{COP})_A}{\gamma} - 1 \right] - f_{AP} \tag{6.48}$$

$$(\text{DCFRR}) = \frac{c_i y}{\overline{C}_{FC}} [(\text{COP})_A - 1] - f_{AP} \tag{6.49}$$

Equations (6.44)–(6.49) neglect the cost of maintenance which is likely to be small in the case of an electrically driven mechanical vapour compression heat pump.

Here is an example calculation. Assume the following values: $y = 8000$ operating hours per year; $\overline{C}_{FC} = \text{US\$800 kW}^{-1}$; $f_{AP} = 0.2$; $c_i = \text{US\$0.05 kW}^{-1}\text{h}^{-1}$. Substitute these values into equation (6.48) to give

$$(\text{DCFRR}) = 0.5 \left(\frac{5}{\gamma} - 1 \right) - 0.2$$

which is the fractional value of the discounted cash flow rate of return for the assumed parameters.

For various values of γ the ratio of the unit cost of the input energy to the unit cost of the base heat supply, the corresponding values of the (DCFRR) and the payback period (PBP) are listed in Table 6.2.

The calculations illustrate the critically sensitive nature of γ. In the case of a gas engine driven heat pump replacing gas heating, the value of $\gamma = 1$.

The most common case is to consider the possible replacement of steam heating with a unit cost of c_{st} in US\$ kWh^{-1} by a mechanical vapour compression heat pump driven by electricity with a unit cost of c_e in US\$ kWh^{-1}. The value of the ratio c_e/c_{st} is dependent on the cost of producing the steam. The greater the cost of producing the steam, the more attractive it becomes to replace the steam heating with the heat produced by an electrically driven mechanically vapour compression heat pump.

Table 6.2

γ	(DCFRR)	(PBP) in years
1.0	1.8	0.55
1.5	0.97	1.03
2.0	0.55	1.80
2.5	0.30	3.30
3.0	0.13	7.70

Although engine driven heat pumps can have attractively small values of the ratio γ, they currently have relatively high capital and maintenance costs compared with electrically driven systems.

6.5   Economics of Heat Driven Absorption Heat Pump Systems

The economics of absorption heat pumps are largely determined by the capital cost of the equipment. An insignificant amount of mechanical energy is required for the pumps. Approximately 95% of the energy input is low-grade heat which, in most cases, can be supplied at little or no cost.

Consider the following options for supplying heat to a particular process:

- heating by a gas boiler;
- heating by a gas driven heat pump.

Let the estimated fixed capital costs of a gas driven heat pump and a gas boiler be C_{HP} and C_B respectively.

Since $C_{HP} > C_B$ and $A_{HP} < A_B$, the payback period in years for the heat pump compared with the gas boiler will be

$$(\text{PBP}) = \frac{C_{HP} - C_B}{A_B - A_{HP}} \tag{6.50}$$

The corresponding approximate discounted cash flow rate of return

$$(\text{DCFRR}) = \frac{A_B - A_{HP}}{C_{HP} - C_B} \tag{6.51}$$

6.6   Economics of Heat Driven Absorption Heat Pump Assisted Water Purification

Although the economics of absorption heat pumps are largely determined by the capital cost of the equipment, other cost elements must be taken into account in order to obtain realistically detailed cost estimates.

6.6.1 The capital element of the production cost

Consider an absorption heat pump providing the heat required to operate a water purification plant which produces M kg h^{-1} of pure water operating for y hours per year. Let the fixed capital cost of the plant be C_{FC} in US\$. The annual cost A in US\$ to borrow the money is given by the equation

$$A = C_{FC}f_{AP} \qquad (6.52)$$

where f_{AP} is the dimensionless annuity present worth factor given by the equation

$$f_{AP} = \frac{i(1+i)^n}{(1+i)^n - 1}$$

where n is the life of the plant in years and i is the fractional annual interest rate on the borrowed money. The capital cost element for the production of pure water c_w in US\$ per kg is given by the equation

$$c_w = \frac{A}{y\,M} \qquad (6.53)$$

which can be written in terms of equation (6.52) as

$$c_w = \frac{C_{FC}f_{AP}}{y\,M} \qquad (6.54)$$

The fixed capital cost C_{FC1} for a plant with a production rate M_1 can be scaled up to give an estimate of the fixed capital cost C_{FC2} for a plant with a higher production rate M_2 using the seven-tenths power law [Holland *et al.*, 1983] given by the equation

$$\frac{C_{FC2}}{C_{FC1}} = \left(\frac{M_2}{M_1}\right)^{0.7} \qquad (6.29)$$

Based on a small plant of fixed capital cost $C_{FC} = $ US\$30 000 and a production rate of pure water of $M = 4.5$ kg h^{-1}, the fixed capital cost of plants with 10 and 100 times the production rates are calculated to be US\$150 000 and US\$750 000 respectively. The capital elements of the production costs for three plants with production rates of 4.5, 45 and 450 kg h^{-1} respectively are given in Table 6.3 for the following annual costs of borrowing where the annuity present worth factors are

$$f_{AP} = 0.19925 \text{ for } i = 0.15 \text{ or } 15\%$$

$$f_{AP} = 0.23852 \text{ for } i = 0.20 \text{ or } 20\%$$

$$f_{AP} = 0.28007 \text{ for } i = 0.25 \text{ or } 25\%$$

Table 6.3 Capital element of the production cost for absorption heat pump assisted water purification

Production rate, M	Capital cost, C_{FC}	Capital element of unit production cost			
		$c_w = \frac{C_{FC} f_{AP}}{y M}$	Annual cost of borrowing (US cents per litre of pure water)		
$(kg\,h^{-1})$	(US$)	$(US\$\,kg^{-1})$	*15%*	*20%*	*25%*
4.5	30 000	$0.8333\,f_{AP}$	16.60	19.90	23.30
45	150 000	$0.4176\,f_{AP}$	8.32	9.96	11.70
450	750 000	$0.2083\,f_{AP}$	4.15	4.97	5.83

Basis of calculations: $y = 8\,000\,h\,y^{-1}; n = 10$ years; $C_{FC2}/C_{FC1} = (M_2/M_1)^{0.7}$

6.6.2   *The heat energy element of the production cost*

The price of oil in US dollars per barrel ($0.159\,m^3$) tends to have a dominant effect on energy prices in general. A barrel of oil is normally quoted as having a thermal energy value of 6.12 GJ. Since 1 GJ $= 277.7\,kWh_t$, one barrel of oil has a thermal energy value of $6.12 \times 277.7 = 1699.5 kWh_t$.

An oil price of US$17 per barrel is equivalent to a thermal energy price of 1 US cent per kWh_t or US$2.78 per GJ. The latent heat of vaporisation of water at $100°C$ is $2257\,kJ\,kg^{-1}$ which is equivalent to $0.6269\,kWh_t\,kg^{-1}$. The energy cost of distilled water with oil heating at 1 US cent per kWh_t would be 0.6269 US cents per litre which is very small compared with the capital element of the production cost of pure water listed in Table 6.3.

Therefore, the cost of the heat supplied to the plant has only a minor effect on the overall economic evaluation of absorption heat pump assisted water purification. It is interesting to compared the cost of heating with gas with heating with oil. British Gas quotes their natural gas from the North Sea as having a thermal energy value of $40.3\,MJ\,m^{-3}$. In 1997, British Gas were charging their domestic customers in the UK 1.4 UK pence per kWh_t which is approximately 2.24 US cents per kWh_t for thermal heating. Even with gas heating at this price, the heating cost element is significantly smaller than the capital cost element of the production cost.

6.6.3   *The labour element of the production cost*

The experimental AHPAEPU constructed and installed at the Instituto de Investigaciones Eléctricas (IIE), Cuernavaca, Mexico, is a small compact prototype which can be transported on the back of a lorry or pickup truck for demonstration purposes. To estimate a labour cost per unit of production for this small research unit would not be meaningful. However, the plant referred to in Table 6.3 with a production rate of 450 kg h^{-1} of pure water would require a full time operator.

The labour element of the production cost c_{LC} in US cents per litre of pure water can be calculated from the equation

$$c_{LC} = \frac{100\,c_L}{M} \tag{6.55}$$

where c_L is hourly labour cost in US$ and M is the production rate of pure water in kg h^{-1}. Substituting $c_L = $ US$10 h^{-1} and $M = 450$ kg h^{-1} into equation (6.55) gives a value for the labour element of the production cost of pure water $c_{LC} = 2.22$ US cents per litre.

6.6.4   The depreciation element of the production cost

The values for the capital element of the production cost listed in Table 6.3 also incorporate a method for plant depreciation in addition to the interest charges on the borrowed money. However, it is also interesting to calculate the depreciation excluding the interest charges on the borrowed money. The simplest method of depreciation is the straight line method where the annual depreciation A_D in US$ per year is given by the equation

$$A_D = \frac{C_{FC} - S}{n} \tag{6.56}$$

where S is the scrap value of the plant. For the case in Table 6.3 of the plant with a fixed capital cost, $C_{FC} = $ US$750\,000$ and zero scrap value, the annual cost of depreciation over $n = 10$ years is $A_D = $ US$75\,000$.

For a production rate $M = 450$ kg h^{-1} operating for 8000 h y^{-1} the depreciation element of the production cost of pure water $c_D = 2.08$ US cents per litre. This is also relatively small compared with the interest charges.

6.6.5   Miscellaneous elements of the production cost

These have been listed in detail in the book by Holland *et al.* [1983]. The annual production cost is the sum of the direct production cost and the indirect production cost. The direct production cost includes those expenses incurred directly for the production operation. These expenses arise from the following items: operating labour, operating supervision, utilities, operating maintenance and repairs, operating supplies, royalties, patents and raw materials (including delivery). Although in the case of an effluent purification plant, the raw materials costs will be zero, delivery costs will be incurred.

The indirect production cost includes those expenses incurred as a result of, but not directly from, the production operation. These expenses arise from the following items: payroll overhead, control laboratory, general plant over-head, packaging and storage facilities. Other indirect costs include those expenses which are a function of the capital investment. These expenses arise from the following items: property taxes, rent and insurance.

A rough average figure for the annual cost of maintenance is 6% of the fixed capital cost of the plant C_{FC}. Royalty and patent costs are of the order of 1–5% of the sales price of the product.

Payroll overhead includes the cost of pensions, holidays, sick pay, etc. It is normally equivalent to 10–20% of the operating labour cost. Plant overhead includes the cost of medical, safety, recreational, effluent disposal and warehousing facilities, etc. The plant overhead cost can vary between the equivalent of 50 and 150% of the operating labour cost.

Property taxes or rates depend on location. They may be taken as 2% of the fixed capital cost of the plant C_{FC} in the absence of specific data. The cost of insurance is normally of the order of 1% of the fixed cost of the plant C_{FC}.

6.6.6   *Other factors to consider in economic evaluations*

In addition to the factors already discussed, there are a number of other factors to take into account, such as marketing, advertising and distribution costs, which must be added to the basic production cost before a realistic profit can be estimated. Nevertheless, the production costs for absorption heat pump assisted water purification are very much lower than the sales price of pure water in Mexico which can be as high as 200 US cents per litre which indicates that the process could be highly profitable.

The further development of the process would also provide useful lessons. The benefits of the learning curve were first quantified as early in 1925 by the commander of the Wright-Patterson US Air Force Base. Subsequent studies on aircraft assembly [Andress, 1954] showed that the fourth aircraft required only 80% as much direct labour as the second, the eighth plane only 80% as much as the fourth, the one hundredth only 80% as much as the fiftieth, and so on. Thus, the rate of learning was concluded to be 80% between doubled quantities.

Domestic refrigeration is a good example of the application of this; not only is the prime cost of a domestic refrigerator much lower today (in real terms) than it was when first introduced, but the maintenance required is minimal. Many domestic refrigerators more than 10 years of age have never broken down.

The electric motor is another good example. This is why oil and gas driven mechanical vapour compression heat pumps have, up to the present, not been able to compete with electrically driven units in spite of the fact that engine driven units can have a significantly higher efficiency based on the primary ratio (PER), discussed in Chapter 1.

The learning process should eventually lead to lower capital costs for absorption heat pump systems. However, the learning process should eventually lead not only to cheaper but to more reliable heat pumps with reduced payback periods. Consider a heat pump application with a payback period of three years. If the heat pump could be fabricated for 80% of the current capital cost in five years' time, then the new payback period would be 2.4

years. At this rate of learning, the payback period could be expected to be 1.92 years for a unit installed in ten years' time and 1.54 years for a unit installed in fifteen years' time.

The effect of inflation on future profitability should also be mentioned. One of the most commonly used methods for estimating the profitability of a process is to calculate the discounted cash flow rate of return (DCFRR) on the basis of projected future net annual cash flows. The true value of the (DCFRR) is then obtained by deducting the annual rate of inflation from the (DCFRR) calculated from the estimated values of the annual cash flows.

However, once an absorption heat pump assisted effluent purification plant has been brought into operation, the only element of the production cost which is likely to be affected by inflation is the labour cost. Although it is important to operate highly economic processes in order to generate national wealth, it may be more important, in the long run, to conserve valuable energy resources by adopting more energy-efficient processes which may appear to be rather less economic. In this respect, absorption systems do conserve energy.

Taking a wider view of economics, it should be matter of national concern that there are many millions of people in Mexico without any sufficiently pure drinking water. Many millions suffer from poor health and disease in Mexico as a result of this deprivation. This does, in turn, have an adverse effect on the overall Mexican economy, irrespective of the decisions whether to go ahead or not with individual processes. Unfortunately, as a result of the historically high inflation rates in Mexico in recent years, industrialists tend to be very conservative, which inhibits the introduction of innovative new technologies, even those with attractive payback periods.

6.7 References

Andress, F. J. (1954) The learning curve as a production tool, *Harvard Business Review*, Jan–Feb. 89.

Holland, F.A., Watson, F. A. and Wilkinson, J. K. (1983) *Introduction to Process Economics*, 2nd edn, John Wiley, New York, USA.

Lang, J. H. (1947) Capital cost estimation, *Chemical Engineering*, **54**, 117–21.

Lang, J. H. (1948) Capital cost estimation, *Chemical Engineering*, **55**, 112–13.

7 Alternative purification systems and the future potential for heat pump technology

7.1 Nomenclature

Q heat flow rate [kW] or heat quantity [kWh]
T temperature [°C or K]
$T_{\bar{A}}$ ambient reference temperature [°C or K]
W rate of work [kW] or quantity of work [kWh]

7.2 Introduction

Figure 7.1 presents a schematic summary of the main treatment methods used to produce water with quality standards for various industrial uses. The energy inputs to these various processes range from high-grade energy, such as electricity, to less valuable heat energy delivered by steam or hot water. The lower the temperature of the heat energy, the less its work equivalent or exergy and consequently its economic value. In fact, a substantial amount of waste heat can be regarded as having little or no economic value, even though it could perhaps be usefully used in heat driven heat pump systems.

Heat driven absorption heat pump assisted purification systems have a number of significant advantages over conventional purification systems which in general are more energy efficient in terms of primary energy consumption, kJ kg^{-1} of product.

The first advantage is that the systems can normally be operated on free environmentally clean low-grade heat energy. For example, a heat transformer can take in low grade heat at a temperature of 65°C and deliver it to the purification system at a temperature of 100°C if approximately half the low-grade heat input can be discharged at a temperature of 30°C. A heat transformer assisted purification system is shown schematically in Figure 7.2. The second advantage is that the systems are extremely simple and can be made portable to be transported on the back of a lorry or pickup truck. Furthermore, they have extremely low maintenance and labour requirements and do not require the addition of chemicals. The third advantage is that the systems are capable of producing a very high purity product and can also be adapted to concentrate chemicals.

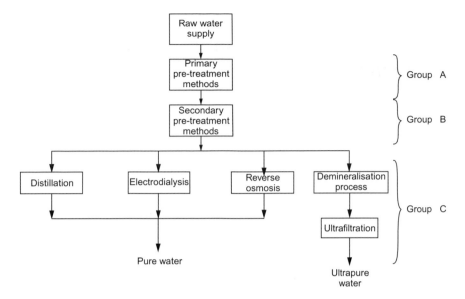

Figure 7.1 Main treatment methods used to produce water with standard quality for various industrial uses.

7.3 Sources of Low-Grade Heat

Low-grade heat, in the temperature range from ambient to the normal boiling point of water, is by far the largest energy resource available to mankind. However, because of the relatively low price and ready availability of more convenient forms of energy, such as non-renewable fossil fuels, comparatively little time, effort and money has so far been given to the efficient use of low grade heat.

Some important sources of low-grade heat are:

- solar heat;
- power stations;
- industrial processes;
- environmental air and water;
- geothermal heat;
- ocean thermal energy.

Low-grade heat can be used in any of the following ways:

- direct use;
- heat exchange and use at a lower temperature;
- conversion to work or electricity;
- recycling at an increased temperature using heat pump technology;

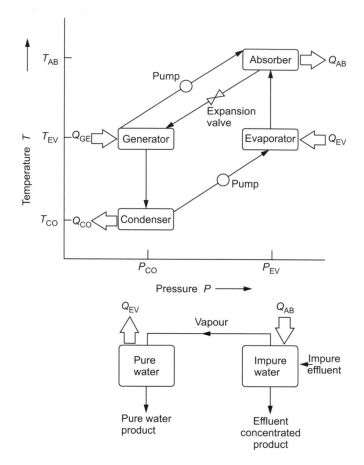

Figure 7.2 Schematic diagram of a heat transformer assisted effluent purification
system.

- provision of cooling using heat pump technology;
- storage for subsequent use.

In the mid-1990s, world demand for primary energy was approximately 170
million barrels per day of oil equivalent, or 390 EJ y^{-1} of which perhaps
approximately 260 EJ y^{-1} was transformed into low grade heat. By contrast
the solar energy falling on the Earth is 5 500 000 EJ y^{-1}, which is more than
14 000 times greater than the current annual world consumption of primary
energy.

7.4   Solar Desalination

Approximately 97% of the Earth's water is in the oceans with an average salt
content of about 3.5% or 35 000 mg litre^{-1}. As the Earth's population

increases and fresh water sources become increasingly polluted, more attention will need to be given to the further development of desalination processes.

The Earth's hydrological cycle which is used to produce rain is a huge solar desalination system. Solar radiation is absorbed as heat into the oceans. This solar heat, in turn, is then used to evaporate water and the water vapour is carried away by the winds, before it is cooled to its dew point when it condenses and precipitates as fresh water rain.

Solar thermal desalination systems can either use solar heat directly to produce distillate in the solar collector or indirectly where the solar collector is coupled to a desalination system.

The use of solar heat in desalination systems, however inefficient, is a far better use of energy than, for example, direct heating using an oil burner with its associated environmental pollution. An oil burner operating at a temperature of 1400°C would have to deliver approximately 2570 kJ of heat to distil 1 kg of water.

The exergy or work equivalent W of a quantity of heat Q is given by the equation

$$W = Q\left(1 - \frac{T_{\bar{A}}}{T}\right) \qquad (7.1)$$

where T is the absolute temperature of the heat source and $T_{\bar{A}}$ is the reference ambient temperature.

In the case of an oil burner operating at a temperature of 1400°C (1673 K) and an ambient reference temperature of 25°C (298 K) the exergy or work equivalent of 1 kJ of heat is 0.822 kJ.

By contrast, in the case of a heat transformer assisted purification system with input heat at a temperature of 65°C (338 K), the exergy or work equivalent of 1 kJ of heat is only 0.118 kJ.

7.5   Fresh Water Resources

The remaining 3% of the Earth's water is fresh water in ground water, lakes and rivers. The rapid growth in population has led to an escalation in the demand for fresh water. Unfortunately, the rapid industrial growth which has accompanied the population explosion has led to increasing amounts of contaminated waste water being discharged into rivers, lakes and underground water reservoirs. This, in turn, will lead to a severe fresh water shortage on a global scale in the relatively near future.

Fresh water in arid zones may contain total dissolved solids of 0.1% or 1000 mg litre^{-1} and 500 mg litre^{-1} is normally taken as the maximum permissible upper limit.

Because of increasing contamination, in the future a larger number of traditional fresh water supply sources will need to be desalinated in the same way as those from arid and semi-arid regions are currently treated.

7.6 Conventional Heat-Consuming Desalting Processes

7.6.1 Solar stills

A conventional solar still, shown schematically in Figure 7.3, uses the greenhouse effect to evaporate salt water which is enclosed in a ∧-shaped glass envelope [Kalorigou, 1997]. The Sun's rays pass through the glass roof and are absorbed by the blackened bottom of the basin to heat the water and increase its vapour pressure. The water vapour is then condensed on the underside of the roof. The condensed distilled water is collected in troughs which conduct it away to the collection system. A number of researchers have used various techniques to improve the performance of solar stills.

7.6.2 Multi-stage flash systems

Multi-stage flash (MSF) systems are the most widely used desalting systems. A schematic diagram of an MSF system is shown in Figure 7.4. It consists of a number of stages. In each stage, condensing steam is used to pre-heat the salt water feed. Commercial MSF installations contain between ten and thirty stages with approximate temperature drops of 2°C per stage.

A disadvantage of MSF systems is that they require very precise pressure levels in each of the different stages, which makes them relatively unsuitable for coupling with solar panels.

7.6.3 Multiple effect boiling systems

A schematic diagram of an multiple effect boiling (MEB) system is shown in Figure 7.5. The heat to the first effect can either be provided by a solar collector or by steam from a conventional boiler. This heat input to the first effect is used to evaporate part of the salt solution. The evaporated water

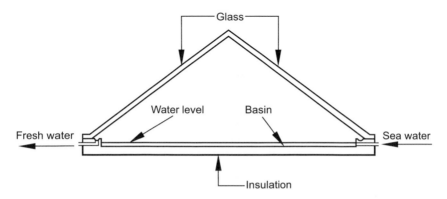

Figure 7.3 Schematic diagram of a conventional solar still.

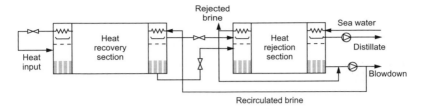

Figure 7.4 Schematic diagram of a multi-stage flash process plant.

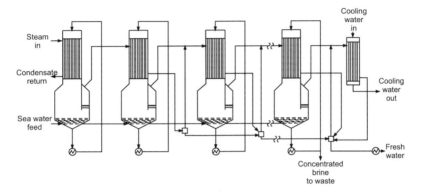

Figure 7.5 Schematic diagram of a long tube multi-effect boiling plant.

from the first effect provides the steam heating for the second effect. As this condenses, it evaporates part of the salt solution in the second effect, and so on. Commercial MEB installations can have up to fourteen effects.

MEB systems are more complex than MSF systems. However, they have the advantage that the operating temperatures and pressures are less critical which makes them rather more suitable for coupling with solar panels.

7.7 Conventional Power-Consuming Desalting Processes

7.7.1 Vapour compression systems

A schematic diagram of a vapour compression (VC) system is shown in Figure 7.6. The heat to the first effect can either be provided by a solar collector or by steam from a conventional boiler. This heat input to the first effect is used to evaporate part of the salt solution. The evaporated water from the first effect is compressed to a higher pressure and is subsequently condensed at a higher temperature. This enhanced heat energy can either be used to heat the first or subsequent stages. VC systems can, with advantage, be combined with MEB systems. However, VC systems have the disadvantage of salt carry-over

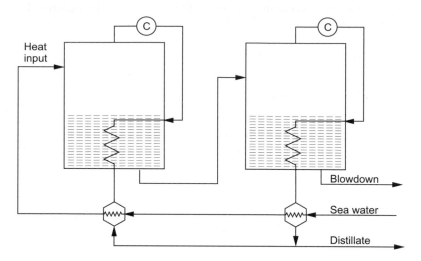

Figure 7.6 Schematic diagram of a vapour compression system.

which leads to corrosion of the compressor blades. They are also limited in size by the availability of suitable compressors.

7.7.2   *Reverse osmosis systems*

A schematic diagram of a reverse osmosis (RO) system is shown in Figure 7.7. When salt water and fresh water are separated by a semi-permeable membrane, the fresh water has a tendency to flow through the semi-permeable membrane to dilute the salt water under the influence of osmotic pressure until equilibrium has been reached. However, if a greater pressure is applied in the reverse direction, relatively pure water will flow through the semi-permeable

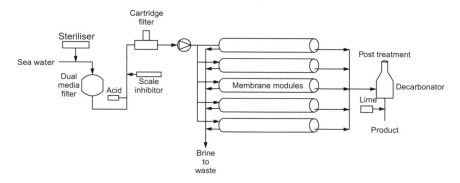

Figure 7.7 Schematic diagram of a reverse osmosis plant.

membrane leaving a concentrated salt solution behind [Houghton, 1989]. Power consumption is proportional to the amount of dissolved solids.

The membranes have the disadvantage that they are susceptible to bacterial fouling when used with certain classes of effluent [Bourgeois, 1982].

7.7.3 Electrodialysis systems

In an electrodialysis (ED) system, an electrical potential difference (PD) is applied across alternatively spaced membranes. Salt water contains positively charged sodium ions and negatively charged chlorine ions. The positively charged sodium cations will move towards the negatively charged electrode and the negatively charged anions will move towards the positively charged electrode when a PD is applied. If special cation-permeable and anion-permeable membranes are placed between the two electrodes, the space between the membranes will gradually become depleted of salts. The energy requirements for electrodialysis are proportional to the salt content of the water [Meller, 1984].

However, ED systems have the further disadvantage that the low conductivity of pure water greatly increases the energy requirements which puts a severe limitation on the purities which can be obtained by this method.

7.7.4 Freezing systems

When a salt solution is frozen, crystals of pure water nucleate and grow in a solution with a progressively increasing concentration of salt. In a conventional freezing system, liquid butane is evaporated in direct contact with the salt water to remove the latent heat of crystallisation. The slurry of ice and concentrated salt water is then pumped to a washer where the slurry rises and the ice crystals are compacted to form a porous bed, which is subsequently removed via a mechanical scraper. The butane vapour is then compressed until it condenses on the ice which is subsequently melted by the heat of condensation of the butane. Finally, the condensed butane and the product water are separated in a decanting unit.

7.8 Pretreatment Methods

Water purification methods, such as reverse osmosis and ultrafiltration, require that the water entering the process should already have a certain degree of purity which is normally achieved by pretreatment. This pre-treatment can include biological control, chemical treatment, adsorption, filtration, chlorination, clarification and ion exchange (Figure 7.8 and Table 7.1). The impurities removed by these processes can cause fouling, scaling and deterioration of the media used in upstream purification processes. The choice of method depends not only on the required purity of the product but also on the initial conditions of the water and the costs of the process.

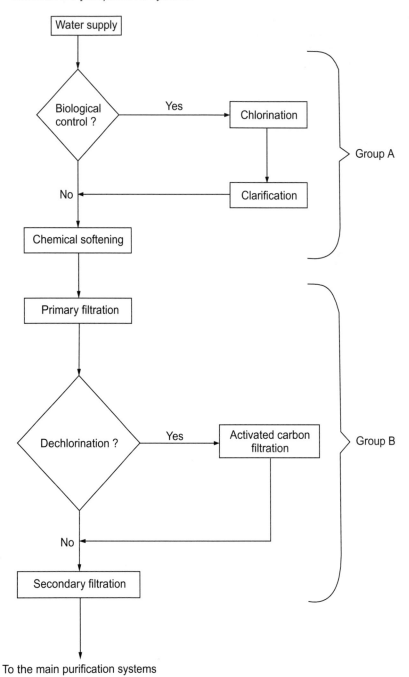

Figure 7.8 Primary and secondary pretreatment processes used as previous control
        purification in the main water purification systems.

Table 7.1 Typical raw water analyses and capabilities for some primary and secondary pretreatment processes [modified from Drew Chemical Corporation (1979)]

Inorganic components	Raw water (mg l⁻¹)	After clarification and filtration (mg l⁻¹)	Percentage removal	After cold lime softening and filtration (mg l⁻¹)	Percentage removal	After clarification, filtration and sodium-cation exchange softening (mg l⁻¹)	Percentage removal	After clarification, filtration and demineralisation (mg l⁻¹)	Percentage removal
Calcium	51.5	51.5	0.0	38.7	24.8	1.0	98.0	0.0	100.0
Magnesium	19.5	19.5	0.0	17.5	10.2	1.0	94.9	0.0	100.0
Sodium	18.6	18.6	0.0	18.6	0.0	87.6*	0.0	1.0	94.6
Potassium	1.8	1.8	0.0	1.8	0.0	1.8	0.0	0.0	100.0
Bicarbonate	56.8	47.8	15.8	0.0	100.0	47.8	15.8	0.0	100.0
Carbonate	0.0	0.0	–	33.0*	–	0.0	–	0.0	–
Hydroxide	0.0	0.0	–	0.0	–	0.0	–	0.0	–
Sulphate	21.8	30.8*	0.0	30.8*	–	30.8*	–	0.0	100.0
Chloride	12.0	12.0	0.0	12.0	0.0	12.0	0.0	0.0	100.0
Nitrate	0.8	0.8	0.0	0.8	0.0	0.8	0.0	0.0	100.0
Iron	0.2	0.0	100.0	0.0	100.0	0.0	100.0	0.0	100.0
Silica	9.0	9.0	0.0	9.0	0.0	9.0	0.0	0.01	99.8
Turbidity (units)	100.0	0–2	98.0	0–2	98.0	0.0	100.0	0.0	100.0
pH	6.5–7.5	6.0–8.0	–	9.0–11.0	–	6.0–8.0	–	7.0–9.0	–

* Represents a re-concentration process on the original raw water composition

Chlorination is capable of controlling biological contamination and the oxidation of heavy metals [White, 1972]. Suspended solids, colloidal substances and finely divided particles can be removed through coagulation, flocculation and sedimentation [ASTM, 1988].

Further pre-treatment processes include chemical softening to reduce hardness, alkalinity, silica and other dissolved solids. An ion exchange process can be used to remove anionic or cationic dissolved solids, colloidal substances and organic matter.

7.9   Comparison of the Water Purities obtained with the Various Desalting Systems

A comparison of desalination systems is dependent on the end use of the purified water since it is not always appropriate to produce highly purified water. This is not only because a particular use may not require such a high purity but, for example, in the case of potable water, highly purified water is unsuitable. In particular, drinking demineralized water can be dangerous. In addition, most systems will require a combination of processes to reach the desired purity. Secondary processes, such as clarification, filtration and cold lime softening, rarely provide sufficiently pure water by themselves. However, these processes are frequently employed to improve the water purity and working life of more expensive systems such as reverse osmosis (RO), demineralisation (DI), electrodialysis (ED), evaporation (EV) and ultrafiltration (UF).

Reverse osmosis and demineralisation have high inorganic component removal efficiencies [Powell Associates, 1990] (Table 7.2). However, both these systems require careful control of the initial water quality to achieve acceptable working lifetimes for the membranes and resins respectively. Ultrafiltration is not suitable for salt removal. However, it does achieve high efficiencies (90–98%) when used for the removal of high molecular weight and colloidal substances [Mercado *et al.*, 1985].

All those systems based on simple distillation (solar stills, multi-stage flash, multiple-effect boiling, vapour compression and heat pump systems) can in principle achieve very high water purities. However, these purity levels can be adversely affected due to carry over of drops or particles from the concentrated waste stream. For potable water production, this may not be important since the final product has to contain an adequate level of dissolved salts. In the case of water sources containing toxic substances, it is necessary to pay careful attention to avoiding carry over.

Distillation based systems can be designed to purify water containing very high levels of dissolved solids and also high levels of suspended solids (Table 7.3). However, under these circumstances, operating efficiencies can be reduced due to high levels of fouling on heat transfer surfaces. One way of ameliorating this heat transfer problem is the use of direct contact heat exchange between an immiscible heat transport fluid and the water to be evaporated.

Table 7.2 Raw water analyses and removal capabilities of a reverse osmosis (RO) system [modified from Desal Co (1997)]

Inorganic components	Raw water $(mg\,l^{-1})$	After RO system $(mg\,l^{-1})$	Percentage removal
Aluminium	5.0	0.02	99.6
Arsenic	42.0	< 0.005	99.9
Barium	74.0	< 0.1	> 99.8
Cadmium	0.2	< 0.005	> 99.5
Chromium	31 900.0	0.01	> 99.9
Cobalt	49.4	< 0.05	> 99.9
Copper	50.0	0.05	99.9
Iron	23.0	0.02	99.9
Lead	0.7	< 0.05	> 93.2
Magnesium	82.1	0.02	99.9
Manganese	1.0	0.01	98.9
Mercury	20.0	< 0.0005	> 99.9
Molybdenum	94.4	< 1.0	> 98.9
Nickel	47.7	< 0.04	> 99.9
Selenium	73.2	< 0.005	> 99.9
Silver	0.1	< 0.01	> 83.0
Sodium	60 000.0	2.4	> 99.9
Vanadium	24.0	< 0.1	99.6
Zinc	74.0	< 0.005	> 99.9
Calcium	276.0	0.1	> 99.9
Chloride	5100.0	0.1	> 99.9
Fluoride	42.2	0.04	99.9
Nitrate	140.0	< 0.4	> 99.7
Nitrite	0.1	< 0.001	> 99.1
Phosphate	14 000.0	0.6	> 99.9
Sulphate	115.0	< 0.001	> 99.9

Table 7.3 Raw and treated water for a simple distillation based heat pump assisted geothermal brine purification system [modified from Siqueiros *et al.* (1995)]

Inorganic components	Geothermal brine supply $(mg\,l^{-1})$	Distillate $(mg\,l^{-1})$	Percentage removal
Silica	1006	2.4	99.8
Chlorides	4146	8.1	99.8
Boron	276	2.1	99.2
Sulphates	16	N.D.	100.0
Sodium	1904	N.D.	100.0
Potassium	491	N.D.	100.0
Lithium	23	N.D.	100.0
Calcium	16	N.D.	100.0
Arsenic	32	N.D.	100.0

N.D. = not detected

To design a water treatment process correctly, a comprehensive economic analysis is necessary. This study must take into account the water purity required, the proposed process and the water production rate. Realistic costs for materials, such as ion exchange resins, membranes and chemicals, must also take into account forecasted cost changes during the plant lifetime.

7.10   Potable Water Requirements

Worldwide, the demand for potable water is increasing, especially in the largest cities. This demand is to satisfy both domestic and industrial requirements. In remote zones, arid and semi-arid areas, potable water supplies are scarce and urgently need improving. Minimum standards for potable water have been established by the World Health Organization [1991, 1992] (Table 7.4). These include physical, chemical and biological specifications.

In arid and semi-arid zones, the World Health Organization limits for salt content are frequently exceeded by a wide margin where only brackish or sea water is available. In tropical and subtropical regions, very often the main problem is with pathogenic bacteria. For example, the Mexican Health Secretariat has produced a standard which differs from the international standards (Table 7.5) [Secretaria de Salud, 1996].

7.11   Industrial Water Requirements

There is a wide range of quality requirements for industrial water which depend on its use. Equally, there is a wide choice of processes available to the

Table 7.4 Maximum concentration limits of the chemical substances in potable water [modified from World Health Organization (1991, 1992)]

Chemical substance	Limits $(mg\,l^{-1})$
Carbon dioxide	20
Carbonates of sodium and potassium	150
Chlorides	250
Chlorine (free)	1.0
Copper	3.0
Detergents	1.0
Fluorides	1.5
Iron	0.3
Lead	0.1
Magnesium	125
Nitrates	10
Phenols	0.001
Sulphates	250
Zinc	15
Total solids in suspension	500
NaCl	250

Table 7.5 Allowable chemical limits for potable water in Mexico [modified from Secretaria de Salud (1996)]

Chemical substance	Limits $(mg\,l^{-1})$
Aluminium	0.20
Arsenic	0.05
Barium	0.70
Cadmium	0.005
Cyanide (as CN^-)	0.07
Free remaining chloride	0.2–1.50
Chloride (as Cl^-)	250.00
Copper	2.00
Chromium	0.05
Hardness (as $CaCO_3$)	500.00
Phenols or phenolics compounds	0.001
Iron	0.30
Fluorides (as F^-)	1.50
Manganese	0.15
Mercurium	0.001
Nitrates (as N)	10.00
Nitriles (as N)	0.05
Ammoniacal nitrogen (as N)	0.50
pH	6.5–8.5
Pesticide: aldrin and dieldrin	0.03
DDT (total isomers)	1.00
Gamma-HCH (lindane)	2.00
Hexachlorobenzene	0.01
Heptachloride and heptachloride's epoxide	0.03
Metoxichloride	20.00
2-4 D	50.00
Lead	0.025
Sodium	200.00
Total solids dissolved	1000.00
Sulphates (as SO_4^{2-})	400.00
Substances active to methylene blue	0.50
Total trihalomethanes	0.20
Zinc	5.00

industrial water treatment designer (Table 7.6) [Drew Chemical Co., 1979]. Figure 7.1 shows a simplified diagram of the most common pre-treatment and treatment processes for producing clear, pure or ultrapure water. These are divided into three groups (A, B and C).

Group A includes pre-treatment methods such as aeration, clarification and lime softening (hot and cold). These methods are used either as the first step for higher quality water production or for cooling water circuit makeup supplies.

Group B encompasses secondary methods such as filtration and adsorption. These processes remove suspended solids usually leaving water suitable for paper cooling, rinsing and potable water for beverages.

Table 7.6 Standards for reagent water ASTMD 1193–77 [modified from DeSilva (1983)]

Parameters	Water type			
	I	II	III	IV
Electrical conductivity (μS cm^{-1} at 25°C)	< 0.06	< 1.0	< 1.0	< 5.0
Electrical resistivity (S cm^{-1} at 25°C)	> 16.67	> 1.0	> 1.0	> 0.2
Soluble silica (μg l^{-1})	< 10	< 10	10	> 10
Colour retention time of KMnO$_4$ (minute)	< 60	< 60	< 10	< 10
Total matter (mg l^{-1})	< 0.1	< 0.1	< 1.0	< 2.0
pH (at 25 °C)	< 7.0	< 7.0	6.2–7.5	5.0–8.0

Group C contains processes that remove or alter dissolved substances by means other than precipitation. Included in this category are distillation, electrodialysis, reverse osmosis, ultrafiltration and demineralisation [Powell Associates, 1990]. The resultant water can be pure or ultrapure. The latter is required for high-pressure steam in electrical power generation, pharmaceutical processes and in the manufacture of semiconductors.

Tables 7.7 and 7.8 present some American Society of Materials Standards (ASTM) of water purity for use in semiconductor and electronics manufacture. Impurities can produce deposits, corrosion and microbiological fouling.

Table 7.7 Chemical requirements for semiconductor-grade purified water set by the Semiconductor Equipment and Materials International (SEMI) [modified from Powell Associates (1990)]

Chemical substance	Limits
Residue (mg l^{-1})	0.10
Total organic carbon (μg l^{-1})	< 20.0
Particulates per litre	< 500
Bacteria per 100 ml	0
Dissolved silica (μg l^{-1})	< 3.0
Resistivity (S cm^{-1})	18.3
Na (μg l^{-1})	0.05
K (μg l^{-1})	0.10
NH$_4$ (μg l^{-1})	< 0.30
Zn (μg l^{-1})	0.02
Fe (μg l^{-1})	0.02
Cu (μg l^{-1})	0.02
Cr (μg l^{-1})	0.02
Mn (μg l^{-1})	0.05
Al (μg l^{-1})	0.2
Cl (μg l^{-1})	0.05
Br (μg l^{-1})	< 0.10
NO$_2$ (μg l^{-1})	0.05
PO$_4$ (μg l^{-1})	< 0.20
SO$_4$ (μg l^{-1})	0.05

Table 7.8 ASTM chemical requirements for electronic-grade water [modified from DeSilva, (1996)]

Chemical substance	Water type			
	I	*II*	*III*	*IV*
Resistivity ($S\,cm^{-1}$)	18	15	2	0.6
Silica ($\mu g\,l^{-1}$)	5	50	100	1000
Particulates ($\mu g\,l^{-1}$)	2	5	100	500
Microorganisms per ml	1	10	50	100
Total organic carbon ($\mu g\,l^{-1}$)	50	200	1000	1000
Copper ($\mu g\,l^{-1}$)	1	5	50	500
Chloride ($\mu g\,l^{-1}$)	2	10	100	1000
Potassium ($\mu g\,l^{-1}$)	2	10	100	500
Sodium ($\mu g\,l^{-1}$)	1	10	200	1000
Zinc ($\mu g\,l^{-1}$)	5	20	200	500
Residue solids ($\mu g\,l^{-1}$)	10	50	500	2000

7.12   Future Developments

The demand for purified water can be divided into two basic categories which require very different levels of purity. The first category includes very pure water for industrial processes, such as that required in the electronics industry, and the second category includes potable water required for human consumption.

The demand for high-purity industrial process water is likely to increase with the rapidly increasing use of microscopic technologies ranging from electronics to micromechanics in mass-produced products. Currently combined methods of purification are commonly used, and these almost always include a distillation stage.

In the potabilisation of water, reverse osmosis has been highly successful in extending brackish water supplies. However, in the case of the desalination of sea water, particularly of warm subtropical sea water, reverse osmosis membranes have a diminished working life. Membrane technology will no doubt improve in the future to extend the working life of membranes under adverse conditions and to enable lower operating pressure differences to be used. However, for conditions where potable water has to be obtained from warm sea water with heavy loads of suspended solids and bacteria, for example, in the Middle East, distillation is likely to be the best option.

For both of the end uses described above, improved distillation technologies would reduce costs and this is likely to happen.

Capital costs will be reduced through better materials technology and improved thermal design. For large-scale potable water systems, distillation towers are vertical rather than the horizontal arrangements commonly used in multiple-effect boiling plants or multi-stage flash systems. This results in a smaller land requirement which in some cases is a significant saving.

The operating costs of future distillation systems will be reduced through better thermodynamic design, such as the use of heat pump transformers (Figure 7.2), and combined heat pump and heat transformer cycles. These cycles have been demonstrated to yield coefficients of performance in excess of four in small-scale laboratory equipment [Ziegler, 1993].

Water pinch technology can complement the use of water purification heat-pump systems. This technology reduces raw-water consumption and waste water in industry by maximising the re-use of water within the plant [Tripathi, 1996]. Mass exchange integration or pinch in mass transfer (water pinch) is a tool that determines the minimum water flow rate required for a plant [El-Halwagi and Manousouthakis, 1989]. Water pinch uses water purity and flow rate instead of temperature and enthalpy as the process parameters. Wang and Smith [1994] considered water pinch as a contamination-transfer problem from process streams to water streams (Figure 7.9). Figure 7.9 shows the contaminant flows from the rich process stream relative to the water stream. The mass transfer driving force is represented by the gap between the process and water stream profiles. The main strategy is to seek opportunities to use the outlet water from one operation to satisfy the water

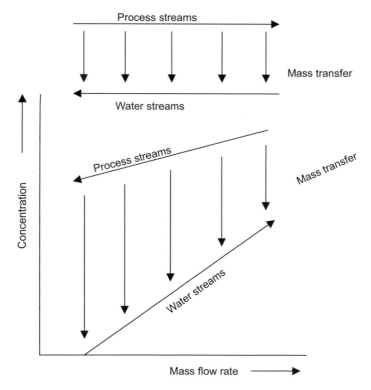

Figure 7.9 Mass transfer concepts used in water pinch technology.

demand of another or the same operation. This methodology means that water needs pH adjustment, membrane separation, ion exchange or sour water stripping, prior to re-use [Dhole *et al.*, 1996]. A real-life successful application of water pinch technology in Wales reduced both raw-water demand and waste water generation [TCE Excellence in Safety and Environment Awards, 1995].

Despite these improvements in technologies for water purification, future demands for water would be better and more cheaply met by superior husbandry of existing resources. This will include much more recycling of water for industrial processes with recovery of contaminants. For drinking water and domestic water supplies, improved distribution systems with better maintenance and system control will reduce the enormous wastage that occurs at present. These improvements will be driven by a change in the way that water is supplied.

At present, water supply in many parts of the world is seen as a public service that delivers a product that is abundant and worthy of subsidy for all. However, clean water is no longer so abundant in parts of the world where it once was and in many places it never was. Some countries, such as France, have a long history of private companies being responsible for water supply and distribution. The trend to the privatisation of water supply is currently on the increase worldwide.

There are now many examples of point of consumption prices being raised to reflect the real cost of supply. This has led to industrial processes that were previously cooled by once-through systems, being modified to include cooling towers and, in some cases, dry cooling towers. In cities such as Mexico City, the size of cisterns on sale for water closets is limited and substitution programmes have replaced older higher capacity units.

Recycling of water requires purification and care that trace substances do not increase to dangerous concentrations. One of the best ways of both avoiding trace substance concentration and facilitating the recovery of valuable substances from industrial water is by distillation. Heat pump and heat pump transformer assisted distillation can make this process highly efficient and economically attractive. If heat is available at an adequate temperature either as a by product of an industrial process or as a renewable resource, such as geothermal or solar energy, absorption heat pump systems can be operated very inexpensively.

One of the most promising potential applications for heat pump assisted distillation technology is in the production of salt-free water from sea water on ocean-based oil rigs. Although an oil rig at sea has no ready access to land-based utilities, such as clean water and electricity, it does have immediate access to a plentiful supply of heat energy from the combustion of hydrocarbons which can be used to drive an absorption heat pump assisted purification system to provide purified salt-free water from sea water. This salt-free water is required for use in various processes in addition to that needed for human consumption.

The heat pump technology described in this book, which has been developed in Mexico since the late 1970s in conjunction with the University of Salford, UK, provides a sound basis for the design of small or large scale water purification plants to be located on oil rigs at sea.

It is interesting to note that if the very approximate cost data for the very small absorption heat pump assisted purification units considered in Table 6.3 (p. 136) were to be scaled up to a production rate of 4 500 000 kgh^{-1}, the annual cost of borrowing could be reduced to 0.2 US cents per litre of pure water with a 10% annual borrowing rate. This production rate would provide a city of 450 000 people with 240 litres per person per day of pure water. With more sophisticated design, involving compact heat exchangers and more accurate and reliable cost data, it may be possible to reduce the cost to as little as 0.1 US cents per litre.

7.13 References

ASTM (1988) Standard Practice for coagulation-flocculation for test of water (D2035-80), *Annual Book of American Society for Testing and Materials Standards (ASTM): Water and Environmental Technology*, **11** (01), Philadelphia, Pennsylvania, USA.

Bourgeois, H. S. (1982) Status of reverse osmosis vs. ion exchange for petroleum/ petrochemical utilities, *Industrial Water Engineering*, **21**(2), 44–8.

Desal Co. (1997) Reverse osmosis products, http://www.desalco.bm/.

DeSilva, F. (1996) Tips for process water, *Chemical Engineering*, August, 72–82.

Dhole, V. R., Ramchandani, N., Tanish, R. A., and Wasilewski, M. (1996) Make your process wastewater pay for itself, *Chemical Engineering*, **103**(1), 100–3.

Drew Chemical Corporation (1979) Drew principles of industrial water treatment, Drew Chemical Co., One Drew Chemical Plaza, Boonton, New Jersey, USA.

El-Halwagi, M. M. and Manousouthakis, V. (1989) Synthesis of mass-exchange networks, *AIChE Journal*, **35**(8), 1233–44.

Houghton, E. J. (1989) Effective solutions to industrial wastewater treatment, in *Proceedings of the 1989 SME International Conference and Exposition*, May 1–4, Detroit, Michigan, USA, pp. 1–11.

Kalogirou, S. (1997) Survey of solar desalination systems and system selection, *Energy*, **22** (1), 69–81.

Meller, F. H. (1984) *Electrodialysis-Electrodialysis Reversal Technology*, Ionics Inc., Watertown, Mass., USA.

Mercado, S., Santoyo, E., Gamino, H. and Lopez-Rubalcaba, H. (1985) Colloidal silica removal from geothermal waters using ultrafiltration systems, *Geothermal Resources Council, Transactions*, **9** (2), 263–7.

S. T. Powell Associates (1990) *Guidelines for Makeup Water Treatment*, Electric Power Research Institute, EPRI GS-6699, Final Report, Palo Alto, California, USA.

Secretaria de Salud (1996) *Salud ambiental, agua para uso y consumo humano- Limites permisibles de calidad y tratamientos a que debe someterse el agua para su potabilizacion*, Norma Oficial Mexicana NOM-127-SSA1-1994, Diario Oficial de la Federacion de Mexico, Enero 18, pp. 41–6.

Siqueiros, J., Heard, C. L. and Holland, F. A. (1995) The commissioning of an integrated heat pump-assisted geothermal brine purification system, *Heat Recovery Systems and CHP*, **15**(7), 655–64.

TCE Excellence in Safety and Environment Awards, (1995) *The Chemical Engineer*, 589.

Tripathi, P. (1996) Pinch technology reduces wastewater, *Chemical Engineering*, **103** (11), 87–90.

Wang, Y. P. and Smith, R. (1994) Wastewater minimisation, *Chemical Engineering Science*, **49**(7), 981–1006.

White, C. G. (1972) *Handbook of Chlorination*, Van Nostrand Reinhold, New York, USA.

World Health Organization (1991) *Revision of the WHO Guidelines for Drinking-water Quality*, IPS, International Programme on Chemical Safety, United Nations Environment Programme, International Labour Organization, World Health Organization, Geneva, Switzerland.

World Health Organization (1992) *WHO Guidelines for Drinking-water Quality, Recommendations*, **1**, World Health Organization, Geneva, Switzerland.

Ziegler, F. (1993) *Absorption refrigeration, heat pump, and heat transformer cycles*, Seminario Recuperación de Calor Industrial por Medio de Bombas de Calor, 17–19 November, Instituto de Investigaciones Eléctricas, Cuernavaca, Morelos, Mexico.

Appendix 1   Derived thermodynamic design data for heat pump systems operating on R718

See *Thermodynamic Design Data for Heat Pump Systems*, Holland, F. A., Watson, F. A. and Devotta, S., Pergamon Press, Oxford, UK (1982).

Chemical name	Water
Chemical formula	H_2O
Molecular weight	18.0
Critical temperature (°C)	373.0
Critical pressure (bar)	221.2
Critical density ($kg\, m^{-3}$)	319.7
Normal boiling point (°C)	100.0
Freezing point (°C)	0.0
Safety group/class	–

Table A1.1 Physical data for R718

T_{CO} (°C)	P_{CO} (bar)	Density ($kg\, m^{-3}$)		PV (bar $m^3 kg^{-1}$)	Latent heat		Enthalpy of saturated vapour ($kJ\ kg^{-1}$)
		Liquid	Vapour		($kJ\ kg^{-1}$)	($MJ\ m^{-3}$) vapour	
0.01	0.006 108	999.80	0.004 847	1.260 16	2501	0.012 12	2601
20	0.023 37	998.20	0.017 29	1.351 65	2454	0.042 43	2637
40	0.073 75	992.16	0.051 15	1.441 84	2406	0.123 07	2674
60	0.199 17	983.19	0.130 2	1.529 72	2358	0.307 01	2709
80	0.473 6	971.82	0.293 4	1.614 18	2308	0.677 17	2743
100	1.013 1	958.31	0.597 7	1.695 00	2257	1.349 01	2776
120	1.985	943.13	1.121	1.771 10	2202	2.468 44	2806
140	3.614	926.10	1.966	1.838 25	2145	4.217 07	2834
160	6.180	907.36	3.258	1.896 87	2082	6.783 16	2858
180	10.027	886.92	5.157	1.944 35	2015	10.391 36	2878
200	15.551	864.68	7.862	1.978 00	1941	15.260 14	2893
220	23.201	840.34	11.62	1.996 64	1859	21.601 58	2902
240	33.480	813.60	16.76	1.997 61	1766	29.598 2	2903
260	46.94	784.01	23.72	1.978 92	1661	39.398 9	2896
280	64.19	750.69	33.19	1.934 02	1542.9	51.208 8	2880
300	85.92	712.45	46.21	1.859 34	1404.3	64.893 0	2849
320	112.90	667.11	64.72	1.744 44	1237.8	80.110 4	2800
340	146.08	610.13	92.76	1.574 82	1027.0	95.264 5	2722
360	186.74	527.98	144.00	1.296 81	719.3	103.579 2	2581

Table A1.2 Theoretical Rankine coefficient of performance $(COP)_R$ for a range of gross temperature lifts $(T_{CO} - T_{EV})$ and condensing temperatures T_{CO} for R718

	T_{CO} (°C)														
	80.0	85.0	90.0	95.0	100.0	105.0	110.0	115.0	120.0	125.0	130.0	135.0	140.0	145.0	150.0
	P_{CO} (bar)														
	0.474	0.578	0.701	0.845	1.013	1.208	1.433	1.691	1.985	2.321	2.701	3.130	3.614	4.155	4.760
$(T_{CO} - T_{EV})$ (°C)	$(COP)_R$ (dimensionless)														
10.0	34.98	35.79	36.30	36.34	36.38	37.66	38.18	38.56	38.77	39.14	39.47	40.50	41.26	41.69	42.12
20.0	17.39	17.55	17.90	18.11	18.26	18.56	18.71	19.13	19.31	19.50	19.63	19.96	20.23	20.42	20.89
30.0	11.49	11.63	11.84	11.92	12.08	12.30	12.42	12.60	12.69	12.91	13.03	13.22	13.35	13.47	13.71
40.0	8.54	8.64	8.80	8.88	9.01	9.12	9.24	9.39	9.47	9.59	9.66	9.84	9.95	10.04	10.17
50.0	6.79	6.88	6.98	7.04	7.15	7.24	7.34	7.43	7.52	7.62	7.69	7.80	7.87	7.98	8.08
60.0	5.60	5.69	5.77	5.84	5.91	5.99	6.07	6.15	6.22	6.29	6.36	6.46	6.52	6.59	6.66
70.0	4.77	4.84	4.90	4.97	5.04	5.11	5.16	5.23	5.29	5.36	5.42	5.48	5.54	5.61	5.67

Table A1.3 Compression ratio (CR) = P_{CO}/P_{EV} for a range of gross temperature lifts ($T_{CO} - T_{EV}$) and condensing temperatures T_{CO} for R718

	T_{CO} (°C)														
	80.0	85.0	90.0	95.0	100.0	105.0	110.0	115.0	120.0	125.0	130.0	135.0	140.0	145.0	150.0
	P_{CO} (bar)														
	0.474	0.578	0.701	0.845	1.013	1.208	1.433	1.691	1.985	2.321	2.701	3.130	3.614	4.155	4.760
($T_{CO} - T_{EV}$) (°C)	(CR) (dimensionless)														
10.0	1.519	1.500	1.480	1.462	1.445	1.429	1.414	1.400	1.386	1.373	1.360	1.349	1.338	1.328	1.317
20.0	2.378	2.311	2.249	2.192	2.139	2.090	2.043	2.000	1.960	1.921	1.885	1.851	1.820	1.790	1.762
30.0	3.839	3.673	3.520	3.379	3.250	3.133	3.025	2.924	2.832	2.746	2.666	2.591	2.523	2.458	2.397
40.0	6.422	6.031	5.684	5.369	5.086	4.830	4.596	4.385	4.192	4.015	3.853	3.703	3.567	3.440	3.323
50.0	11.167	10.283	9.506	8.817	8.213	7.674	7.192	6.759	6.370	6.020	5.703	5.414	5.155	4.917	4.698
60.0	20.263	18.259	16.531	15.032	13.737	12.603	11.614	10.740	9.968	9.280	8.665	8.119	7.631	7.188	6.789
70.0	38.580	33.921	29.997	26.692	23.889	21.486	19.426	17.638	16.096	14.745	13.561	12.514	11.594	10.778	10.051

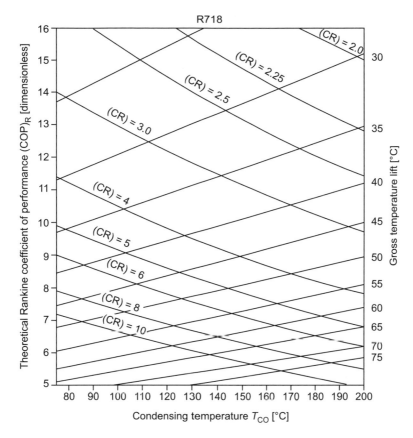

Figure A1.1 Theoretical Rankine coefficient of performance $(COP)_R$ against the condensing temperature T_{CO} for various values of (CR) and gross temperature lift $(T_{CO} - T_{EV})$ for water (R718).

Appendix 2 Derived thermodynamic design data for heat pump systems operating on R12

See *Thermodynamic Design Data for Heat Pump Systems*, Holland, F. A., Watson, F. A. and Devotta, S., Pergamon Press, Oxford, UK (1982).

Chemical name	Dichloro difluoro methane
Chemical formula	CCl_2F_2
Molecular weight	120.9
Critical temperature (°C)	112.0
Critical pressure (bar)	41.15
Critical density, ($kg\,m^{-3}$)	558.0
Normal boiling point (°C)	−29.79
Freezing point, (°C)	−157.7
Safety group/class	1/6

Table A2.1 Physical data for R12

T_{CO} (°C)	P_{CO} (bar)	Density ($kg\,m^{-3}$)		PV (bar $m^3 kg^{-1}$)	Latent heat		Enthalpy of saturated vapour ($kJ\,kg^{-1}$)
		Liquid	Vapour		($kJ\,kg^{-1}$)	($MJ\,m^{-3}$) vapour	
0	3.0861	1396.8	18.055	0.170 93	151.478	2.735	251.478
10	4.2330	1363.8	24.443	0.173 18	146.364	3.578	255.687
20	5.6729	1329.0	32.490	0.174 61	140.908	4.578	259.729
30	7.4490	1292.3	42.540	0.175 10	135.027	5.774	263.567
40	9.6065	1253.1	55.036	0.174 55	128.612	7.078	267.147
50	12.1932	1211.1	70.574	0.172 78	121.513	8.576	270.396
60	15.259	1165.35	90.002	0.169 58	113.520	10.217	273.210
70	18.858	1114.67	114.617	0.164 53	104.325	11.957	275.427

Table A2.2 Theoretical Rankine coefficient of performance (COP)$_R$ for a range of gross temperature lifts ($T_{CO} - T_{EV}$) and condensing temperatures T_{CO} for R12

	T_{CO} (°C)							
	40	45	50	55	60	65	70	75
	P_{CO} (bar)							
	(9.6065)	(10.8431)	(12.1932)	(13.6630)	(15.259)	(16.988)	(18.858)	(20.874)
($T_{CO} - T_{EV}$) (°C)	(COP)$_R$ (*dimensionless*)							
10.0	29.01	29.25	29.45	29.60	29.68	29.71	29.69	29.55
20.0	14.07	14.18	14.26	14.32	14.36	14.37	14.33	14.25
30.0	9.10	9.17	9.22	9.25	9.27	9.27	9.23	9.18
40.0	6.63	6.67	6.71	6.73	6.74	6.73	6.70	6.66
50.0	5.16	5.19	5.21	5.23	5.23	5.22	5.20	5.16
60.0	4.19	4.21	4.23	4.23	4.23	4.22	4.20	4.17
70.0	3.50	3.52	3.53	3.53	3.53	3.52	3.50	3.47

Table A2.3 Compression ratio (CR) = P_{CO}/P_{EV} for a range of gross temperature lifts ($T_{CO} - T_{EV}$) and condensing temperatures T_{CO} for R12

	T_{CO} (°C)							
	40	45	50	55	60	65	70	75
	P_{CO} (bar)							
	(9.6065)	(10.8431)	(12.1932)	(13.6630)	(15.259)	(16.988)	(18.858)	(20.874)
($T_{CO} - T_{EV}$) (°C)	(CR) (*dimensionless*)							
10.0	1.29	1.28	1.27	1.26	1.25	1.24	1.24	1.23
20.0	1.69	1.66	1.64	1.61	1.59	1.57	1.55	1.53
30.0	2.27	2.21	2.15	2.10	2.05	2.00	1.96	1.93
40.0	3.11	2.99	2.88	2.78	2.69	2.61	2.53	2.46
50.0	4.38	4.16	3.95	3.77	3.60	3.46	3.32	3.20
60.0	6.36	5.94	5.56	5.24	4.94	4.69	4.45	4.25
70.0	9.57	8.77	8.08	7.48	6.96	6.51	6.11	5.76

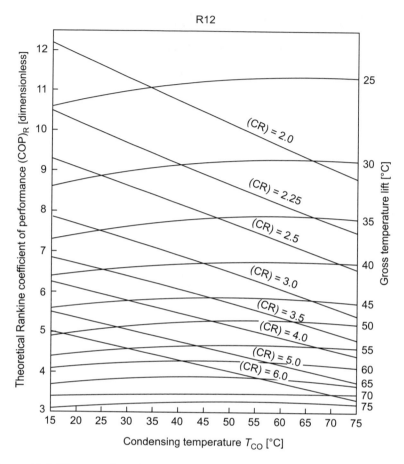

Figure A2.1 Theoretical Rankine coefficient of performance $(COP)_R$ against condensing temperature T_{CO} for various values of (CR) and gross temperature lift $(T_{CO} - T_{EV})$ for (R12).

Appendix 3 Derived thermodynamic design data for heat pump systems operating on R114

See *Thermodynamic Design Data for Heat Pump Systems*, Holland, F. A., Watson, F. A. and Devotta, S., Pergamon Press, Oxford, UK (1982).

Chemical name	Dichloro tetrafluoro ethane
Chemical formula	$CClF_2CClF_2$
Molecular weight	170.9
Critical temperature (°C)	145.7
Critical pressure (bar)	32.63
Critical density ($kg\,m^{-3}$)	582.0
Normal boiling point (°C)	3.78
Freezing point (°C)	−93.9
Safety group/class	1/6

Table A3.1 Physical data for R114

T_{CO} (°C)	P_{CO} (bar)	Density ($kg\,m^{-3}$) Liquid	Density ($kg\,m^{-3}$) Vapour	PV (bar m^3kg^{-1})	Latent heat ($kJ\,kg^{-1}$)	Latent heat ($MJ\,m^{-3}$) vapour	Enthalpy of saturated vapour ($kJ\,kg^{-1}$)
20	1.8047	1471.2	13.521	0.13347	130.853	1.769	250.343
30	2.4968	1440.6	18.406	0.13565	126.731	2.333	256.587
40	3.3723	1408.7	24.570	0.13725	122.220	3.003	262.792
50	4.4577	1375.2	32.254	0.13821	117.343	3.785	268.923
60	5.7802	1340.0	41.755	0.13843	112.116	5.780	274.937
70	7.370	1302.65	53.48	0.13781	105.957	5.667	280.782
80	9.255	1262.6	67.93	0.13624	100.558	6.831	286.388
90	11.471	1219.2	85.93	0.13349	96.079	8.084	291.646
100	14.058	1171.5	108.752	0.12927	86.876	9.448	296.381
110	17.071	1117.6	138.771	0.12302	78.489	10.892	300.270
120	20.575	1054.5	180.997	0.11368	66.793	12.089	302.637

Table A3.2 Theoretical Rankine coefficient of performance $(COP)_R$ for a range of gross temperature lifts $(T_{CO} - T_{EV})$ and condensing temperatures T_{CO} for R114

$(T_{CO} - T_{EV})$ (°C)	T_{CO} (°C)															
	50	55	60	65	70	75	80	85	90	95	100	105	110	115	120	125
P_{CO} (bar)	4.4577	5.0876	5.7802	6.5392	7.370	8.271	9.255	10.317	11.471	12.713	14.058	15.505	17.071	18.753	20.575	22.535
	$(COP)_R$ (dimensionless)															
10.0	30.21	30.51	30.81	31.11	31.36	31.56	31.75	31.91	32.00	32.02	32.01	31.91	31.71	31.31	30.65	29.93
20.0	14.45	14.58	14.71	14.82	14.92	15.01	15.08	15.12	15.15	15.15	15.11	15.03	14.91	14.71	14.41	14.03
30.0	9.22	9.29	9.35	9.41	9.46	9.50	9.53	9.55	9.55	9.53	9.49	9.43	9.33	9.18	8.98	8.73
40.0	6.62	6.66	6.69	6.72	6.75	6.76	6.77	6.77	6.76	6.74	6.70	6.63	6.55	6.43	6.28	6.08
50.0	5.07	5.09	5.11	5.12	5.13	5.13	5.13	5.12	5.10	5.07	5.03	4.97	4.89	4.79	4.66	4.50
60.0	4.06	4.07	4.07	4.07	4.07	4.06	4.05	4.03	4.01	3.97	3.93	3.87	3.80	3.71	3.59	3.45
70.0	3.34	3.34	3.34	3.33	3.32	3.31	3.29	3.27	3.24	3.20	3.15	3.10	3.03	2.94	2.84	2.71

Table A3.3 Compression ratio (CR) $= P_{CO}/P_{EV}$ for a range of gross temperature lifts ($T_{CO} - T_{EV}$) and condensing temperatures T_{CO} for R114

$(T_{CO} - T_{EV})$ (°C)	T_{CO} (°C)															
	50	55	60	65	70	75	80	85	90	95	100	105	110	115	120	125
P_{CO} (bar)	4.4577	5.0876	5.7802	6.5392	7.370	8.241	9.255	10.317	11.471	12.713	14.058	15.505	17.071	18.753	20.575	22.535
(CR) (dimensionless)																
10.0	1.32	1.31	1.30	1.29	1.28	1.26	1.26	1.25	1.24	1.23	1.23	1.22	1.21	1.21	1.21	1.20
20.0	1.79	1.75	1.71	1.68	1.65	1.63	1.60	1.58	1.56	1.54	1.52	1.50	1.49	1.48	1.46	1.45
30.0	2.47	2.39	2.32	2.25	2.19	2.13	2.08	2.03	1.98	1.94	1.91	1.87	1.84	1.82	1.79	1.77
40.0	3.51	3.35	3.20	3.07	2.95	2.84	2.74	2.65	2.57	2.50	2.43	2.37	2.32	2.27	2.22	2.18
50.0	5.14	4.83	4.55	4.30	4.08	3.88	3.71	3.55	3.40	3.27	3.15	3.05	2.95	2.87	2.79	2.72
60.0	7.77	7.18	6.66	6.21	5.80	5.44	5.13	4.84	4.59	4.37	4.17	3.99	3.83	3.69	3.56	3.45
70.0	12.21	11.06	10.08	9.23	8.49	7.85	7.29	6.79	6.36	5.97	5.63	5.33	5.06	4.82	4.62	4.43

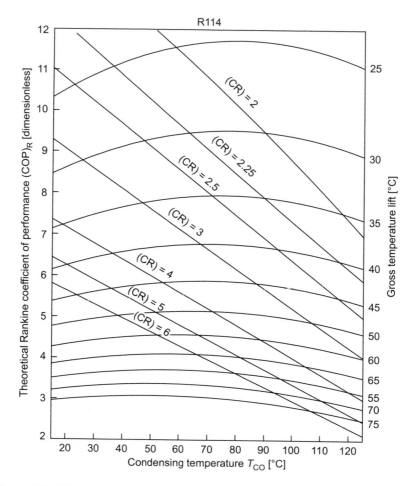

Figure A3.1 Theoretical Rankine coefficient of performance $(COP)_R$ against condensing temperature T_{CO} for various values of (CR) and gross temperature lift $(T_{CO} - T_{EV})$ for R114.

Appendix 4   Derived thermodynamic design data for heat pump systems operating on R717

See *Thermodynamic Design Data for Heat Pump Systems*, Holland, F. A., Watson, F. A. and Devotta, S., Pergamon Press, Oxford, UK (1982).

Chemical name	Ammonia
Chemical formula	NH_3
Molecular weight	17.0
Critical temperature (°C)	132.2
Critical pressure, (bar)	114.3
Critical density (kg m^{-3})	234.5
Normal boiling point, (°C)	−33.33
Freezing point (°C)	−77.7
Safety group/class	2/2

Table A4.1 Physical data for R717

T_{CO} (°C)	P_{CO} (bar)	Density (kg m^{-3}) Liquid	Density (kg m^{-3}) Vapour	PV (bar $m^3 kg^{-1}$)	Latent heat (kJ kg^{-1})	Latent heat (MJ m^{-3}) vapour	Enthalpy of saturated vapour (kJ kg^{-1})
0	4.302 51	638.51	3.46	1.243 50	1262.718	4.3690	1362.718
10	6.160 11	624.71	4.87	1.264 91	1226.611	5.9736	1373.305
20	8.584 02	610.15	6.70	1.281 20	1187.624	7.9571	1381.952
30	11.679 53	595.05	9.05	1.290 56	1145.852	10.3700	1387.994
40	15.564 46	579.48	12.04	1.292 73	1101.638	13.2637	1392.537
50	20.348 45	562.77	15.78	1.289 51	1053.085	16.6177	1395.746
60	26.171 93	545.13	20.54	1.274 19	1001.947	20.5800	1393.761
70	33.159 54	526.27	26.50	1.251 30	943.233	24.9957	1387.890
80	41.474 74	505.59	34.13	1.215 20	877.226	29.9397	1376.297
90	51.286 41	482.92	44.03	1.164 81	802.484	35.3334	1358.929
100	62.688 40	456.99	57.15	1.096 91	714.994	40.8619	1331.036
110	75.915 40	425.81	75.48	1.005 77	607.767	45.8743	1285.680
120	91.024 94	385.39	103.92	0.875 91	450.276	46.7927	1211.123

Table A4.2 Theoretical Rankine coefficient of performance (COP)$_R$ for a range of gross temperature lifts ($T_{CO} - T_{EV}$) and condensing temperatures T_{CO} for R717

	T_{CO} (°C)													
	50	55	60	65	70	75	80	85	90	95	100	105	110	115
P_{CO} (bar)	20.348	23.122	26.172	29.512	33.160	37.137	41.475	46.189	51.286	56.777	62.688	69.059	75.915	83.245
($T_{CO} - T_{EV}$) (°C)	(COP)$_R$ (dimensionless)													
10.0	29.27	32.32	34.89	33.66	31.16	29.92	30.23	31.49	32.39	31.87	30.60	30.60	33.65	38.98
20.0	14.64	15.04	15.74	16.32	16.33	15.73	15.21	15.16	15.40	15.59	15.50	15.36	15.64	16.43
30.0	9.77	9.98	10.18	10.27	10.35	10.46	10.49	10.34	10.16	10.08	10.06	10.13	10.31	10.49
40.0	7.24	7.39	7.53	7.60	7.59	7.57	7.63	7.74	7.78	7.65	7.48	7.43	7.51	7.65
50.0	5.72	5.82	5.92	5.98	6.01	6.00	6.00	6.01	6.06	6.10	6.08	6.00	5.94	5.94
60.0	4.71	4.78	4.85	4.90	4.93	4.94	4.96	4.97	4.98	4.96	4.97	4.99	5.01	4.96
70.0	3.99	4.05	4.11	4.14	4.17	4.18	4.20	4.22	4.23	4.23	4.21	4.19	4.21	4.23

Table A4.3 Compression ratio (CR) = P_{CO}/P_{EV} for a range of gross temperature lifts ($T_{CO} - T_{EV}$) and condensing temperatures T_{CO} for R717

	T_{CO} (°C)													
	50	55	60	65	70	75	80	85	90	95	100	105	110	115
P_{CO} (bar)	20.348	23.122	26.172	29.512	33.160	37.137	41.475	46.189	51.286	56.777	62.688	69.059	75.915	83.245
$(COP)_R$ (dimensionless)														
($T_{CO} - T_{EV}$) (°C)														
10.0	1.307	1.296	1.286	1.276	1.267	1.258	1.251	1.244	1.237	1.229	1.222	1.216	1.211	1.205
20.0	1.742	1.711	1.682	1.655	1.630	1.606	1.585	1.565	1.547	1.529	1.511	1.495	1.480	1.466
30.0	2.371	2.303	2.241	2.183	2.130	2.082	2.038	1.998	1.960	1.924	1.891	1.860	1.830	1.802
40.0	3.303	3.169	3.049	2.939	2.839	2.747	2.665	2.590	2.520	2.456	2.395	2.340	2.289	2.242
50.0	4.729	4.475	4.249	4.045	3.863	3.699	3.551	3.417	3.295	3.183	3.081	2.987	2.901	2.821
60.0	6.982	6.503	6.083	5.712	5.383	5.091	4.832	4.601	4.391	4.200	4.028	3.872	3.731	3.600
70.0	10.678	9.766	8.980	8.300	7.707	7.188	6.733	6.331	5.975	5.655	5.367	5.109	4.877	4.667

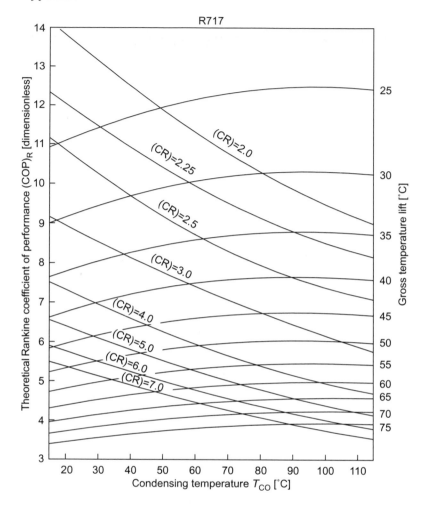

Figure A4.1 Theoretical Rankine coefficient of performance $(COP)_R$ against condensing temperature T_{CO} for various values of (CR) and gross temperature lift $(T_{CO} - T_{EV})$ for R717.

Appendix 5   Compound interest factors

Table A5.1 Compound interest rate of 5 per cent

n (years)	f_i	f_d	f_{AF}	f_{AP}
1	1.050 00	0.952 38	1.000 00	1.050 00
2	1.102 50	0.907 03	0.487 80	0.537 80
3	1.157 62	0.863 84	0.317 21	0.367 21
4	1.215 51	0.822 70	0.232 01	0.282 01
5	1.276 28	0.783 53	0.180 97	0.230 97
6	1.340 10	0.746 22	0.147 02	0.197 02
7	1.407 10	0.710 68	0.122 82	0.172 82
8	1.477 46	0.676 84	0.104 72	0.154 72
9	1.551 33	0.644 61	0.090 69	0.140 69
10	1.628 89	0.613 91	0.079 50	0.129 50
11	1.710 34	0.584 68	0.070 39	0.120 39
12	1.795 86	0.556 84	0.062 83	0.112 83
13	1.885 65	0.530 32	0.056 46	0.106 46
14	1.979 93	0.505 07	0.051 02	0.101 02
15	2.078 93	0.481 02	0.046 34	0.096 34

Table A5.2 Compound interest rate of 10 per cent

n (years)	f_i	f_d	f_{AF}	f_{AP}
1	1.100 00	0.909 09	1.000 00	1.100 00
2	1.210 00	0.826 45	0.476 19	0.576 19
3	1.331 00	0.751 31	0.302 11	0.402 11
4	1.464 10	0.683 01	0.215 47	0.315 47
5	1.610 51	0.620 92	0.163 80	0.263 80
6	1.771 56	0.564 47	0.129 61	0.229 61
7	1.948 72	0.513 16	0.105 41	0.205 41
8	2.143 59	0.466 51	0.087 44	0.187 44
9	2.357 95	0.424 10	0.073 64	0.173 64
10	2.593 74	0.385 54	0.062 75	0.162 75
11	2.853 12	0.350 49	0.053 96	0.153 96
12	3.138 43	0.318 63	0.046 76	0.146 76
13	3.452 27	0.289 66	0.040 78	0.140 78
14	3.797 50	0.263 33	0.035 75	0.135 75
15	4.177 25	0.239 39	0.131 47	0.131 47

Table A5.3 Compound interest rate of 15 per cent

n (years)	f_i	f_d	f_{AF}	f_{AP}
1	1.150 00	0.869 57	1.000 00	1.150 00
2	1.322 50	0.756 14	0.465 12	0.615 12
3	1.520 88	0.657 52	0.287 98	0.437 98
4	1.749 01	0.571 75	0.200 27	0.350 27
5	2.011 36	0.497 18	0.148 32	0.298 32
6	2.313 06	0.432 33	0.114 24	0.264 24
7	2.660 02	0.375 94	0.090 36	0.240 36
8	3.059 02	0.326 90	0.072 85	0.222 85
9	3.517 88	0.284 26	0.050 57	0.209 57
10	4.045 56	0.247 18	0.049 25	0.199 25
11	4.652 39	0.214 94	0.041 07	0.191 07
12	5.350 25	0.186 91	0.034 48	0.184 48
13	6.152 79	0.162 53	0.029 11	0.179 11
14	7.075 71	0.141 33	0.024 69	0.174 69
15	8.137 06	0.122 89	0.021 02	0.171 02

Table A5.4 Compound interest rate of 20 per cent

n (years)	f_i	f_d	f_{AF}	f_{AP}
1	1.200 00	0.833 33	1.000 00	1.200 00
2	1.440 00	0.694 44	0.454 55	0.654 55
3	1.728 00	0.578 70	0.274 73	0.474 73
4	2.073 60	0.482 25	0.186 29	0.386 29
5	2.488 32	0.401 88	0.134 38	0.334 38
6	2.985 98	0.334 90	0.100 71	0.300 71
7	3.583 18	0.279 08	0.077 42	0.277 42
8	4.299 82	0.232 57	0.060 61	0.260 61
9	5.159 78	0.193 81	0.048 08	0.248 08
10	6.191 74	0.161 51	0.038 52	0.238 52
11	7.430 08	0.134 59	0.031 10	0.231 10
12	8.916 10	0.112 16	0.025 26	0.225 26
13	10.699 32	0.093 46	0.020 62	0.220 62
14	12.839 18	0.077 89	0.016 89	0.216 89
15	15.407 02	0.064 91	0.013 88	0.213 88

Table A5.5 Compound interest rate of 25 per cent

n (years)	f_i	f_d	f_{AF}	f_{AP}
1	1.250 00	0.800 00	1.000 00	1.250 00
2	1.562 50	0.640 00	0.444 44	0.694 44
3	1.953 12	0.512 00	0.262 30	0.512 30
4	2.441 41	0.409 60	0.173 44	0.423 44
5	3.051 76	0.327 68	0.121 85	0.371 85
6	3.814 70	0.262 14	0.088 82	0.338 82
7	4.768 37	0.209 72	0.066 34	0.316 34
8	5.950 46	0.167 77	0.050 40	0.300 40
9	7.450 58	0.134 22	0.038 76	0.288 76
10	9.313 23	0.107 37	0.030 07	0.280 07
11	11.641 53	0.085 90	0.023 49	0.273 49
12	14.551 92	0.068 72	0.018 45	0.268 45
13	18.189 89	0.054 98	0.014 54	0.264 54
14	22.737 37	0.043 98	0.011 50	0.261 50
15	28.421 71	0.035 18	0.009 12	0.259 12

Appendix 6 Recommended books on heat pump technology

Alefeld, G. and Radermacher, R. (1993) *Heat Conversion Systems*, CRC Press, Boca Raton, Florida, USA.

Berghmans, J. (1983) *Heat Pumps Fundamentals, Series E*, Applied Science, No. 53, Noordhoff, Leyden, Netherlands.

Brodowicz, K. and Dyakowski, T. (1993) *Heat Pumps*, Butterworth Heinemann, London, UK.

Brombaugh, J. E. (1992) *Heat Pump Fundamentals*, Audel Book Division, Maxwell Macmillan International, New York, USA.

Camatini, E. and Kester, T. (1976) *Heat Pumps and Their Contribution to Energy Conservation, Series E*, Applied Science, No. 15, Noordhoff, Leyden, Netherlands.

Collie, M. J. (1976) *Heat Pump Technology for Saving Energy*, Noyes Publications, Park Ridge, New Jersey, USA.

The European Community's four contractors meetings on heat pumps (1982) The Community's Energy R&D Programme Energy Conservation, Brussels, Belgium.

Electrical Power Research Institute (EPRI) (1988) *Industrial Heat Pump Manual*, EPRI EM 6057, Palo Alto, California, USA.

Electricity Council (1982) *Heat Pumps and Air Conditioning, A Guide to Packaged Systems*, EC 4204, London, UK.

Heap, R. D. (1983) *Heat Pumps*, E&FN Spon, London, UK.

Holland, F. A., Watson, F. A. and Devotta, S. (1982) *Thermodynamic Design Data for Heat Pump Systems*, Pergamon Press, Oxford, UK.

Langley, B. C. (1988) *Heat Pump Technology: Systems Design, Installation and Troubleshooting*, Prentice-Hall, Englewood Cliffs, New Jersey, USA.

Ludwig von Cube, H. and Steimle, F. (1981) *Heat Pump Technology*, Butterworth Heinemann, London, UK.

Moser, F. and Schnitzer, H. (1985) *Heat Pumps in Industry*, Elsevier, Amsterdam, Netherlands.

Radermaher, R., Herold, K. E., Miller, W. Perez-Blanco, H., Ryan, W. and Vleit, G. (1994) *Proceedings of the International Absorption Heat Pump Conference*, AES-31, January 19–21, New Orleans, Louisiana, USA, American Society of Mechanical Engineers (ASME), New York, USA.

Raldow, W. (1982) New working pairs for absorption processes, in *Proceedings of Workshop in Berlin*, April 14–16, Swedish Council for Building Research, Stockholm, Sweden.

Reay, D. A. and Macmichael, D. B. A. (1988) *Heat Pumps Design and Applications*, 2nd edn, Pergamon Press, Oxford, UK.

Sauer, H. J. and Howell, R. H. (1983) *Heat Pump Systems*, John Wiley, New York, USA.

Smith, I. E. (1991) *Applications and Efficiency of Heat Pump Systems in Environmentally Sensitive Times*, Springer-Verlag, Berlin, Germany.

Sumner, J. A. (1976) *An Introduction to Heat Pumps*, Prism Press, Dorchester, UK.

Wurm, J., Kinast, J. A., Roose, T. R. and Staats, W. R. (1990) *Stirling and Vuilleumier Heat Pumps, Design and Applications*, McGraw-Hill, New York, USA.

Appendix 7 Conversion factors for energy equivalents

1 kJ (kilojoule) = 10^3 J = 0.947 Btu = 0.239 kcal
1 MJ (megajoule) = 10^6 J
1 GJ (gigajoule) = 10^9 J = 277.78 kWh
1 TJ (terajoule) = 10^{12} J
1 PJ (petajoule) = 10^{15} J
1 EJ (exajoule) = 10^{18} J = 277.78 TWh

Oil (approximate thermal equivalents)
1 b.o.e. (barrel oil equivalent) = 1742 kWh_t (thermal)
1 billion b.o.e. (barrels oil equivalent) = 6.27 EJ
1 million b.d.o.e. (barrels per day oil equivalent) = 2.29 EJ y^{-1}
100 million t.o.e. (tonnes oil equivalent) = 4.70 EJ
 where 1 tonne = 1000 kg

Coal (approximate thermal equivalent of bituminous coal)
12,000 Btu/lb = 27.912 MJ kg^{-1}
1 tonne = 7.753 kW h_t (thermal)
100 million tonne = 2.79 EJ

Natural gas (approximate thermal equivalents)
1 m^3 = 10.80 kWh_t (thermal)
100 milliard m^3 = 3.89 EJ
where 1 milliard m^3 = 10^9 m^3
10^{12} ft^3 = 1.10 EJ

Index